Computational Intelligence Methods and Applications

Series editors

Sanghamitra Bandyopadhyay, Kolkata, West Bengal, India
Ujjwal Maulik, Kolkata, West Bengal, India
Patrick Siarry, Vitry-sur-Seine, France

The monographs and textbooks in this series explain methods developed in computational intelligence (including evolutionary computing, neural networks, and fuzzy systems), soft computing, statistics, and artificial intelligence, and their applications in domains such as heuristics and optimization; bioinformatics, computational biology, and biomedical engineering; image and signal processing, VLSI, and embedded system design; network design; process engineering; social networking; and data mining.

More information about this series at http://www.springer.com/series/15197

Sourav De · Siddhartha Bhattacharyya
Susanta Chakraborty · Paramartha Dutta

Hybrid Soft Computing for Multilevel Image and Data Segmentation

 Springer

Sourav De
Department of Computer Science
  and Engineering
Cooch Behar Government Engineering
  College
Cooch Behar, West Bengal
India

Susanta Chakraborty
Department of Computer Science
  and Technology
Indian Institute of Engineering Science
  and Technology
Howrah, West Bengal
India

Siddhartha Bhattacharyya
Department of Information Technology
RCC Institute of Information Technology
Kolkata, West Bengal
India

Paramartha Dutta
Department of Computer and System
  Sciences
Visva-Bharati University
Santiniketan, West Bengal
India

ISSN 2510-1765   ISSN 2510-1773   (electronic)
Computational Intelligence Methods and Applications
ISBN 978-3-319-83758-1   ISBN 978-3-319-47524-0   (eBook)
DOI 10.1007/978-3-319-47524-0

Printed on acid-free paper

This Springer imprint is published by Springer Nature
The registered company is Springer International Publishing AG
The registered company address is: Gewerbestrasse 11, 6330 Cham, Switzerland

To my respected parents Satya Narayan De and Tapasi De, loving wife Debolina Ghosh, beloved son Aishik De, sister Soumi De and my in-laws

Sourav De

To my father Late Ajit Kumar Bhattacharyya, mother Late Hashi Bhattacharyya, beloved wife Rashni, father-in-law Mr. Asis Kumar Mukherjee, mother-in-law Mrs. Poly Mukherjee and brother-in-law Rishi Mukherjee

Siddhartha Bhattacharyya

To my father, Late Santosh K. Chakraborty and my beloved students, who can take away the ideas forward

Susanta Chakraborty

To my father, Late Arun Kanti Dutta and mother Mrs. Bandana Dutta

Paramartha Dutta

Preface

Segmentation is targeted to partition an image into distinct regions comprising pixels having similar attributes. In the context of image analysis and interpretation, these partitioned regions should strongly relate to depicted objects or features of interest. Faithful segmentation of an image scene is the first step from low-level image processing transforming a multilevel or colour image into one or more other images to high-level image description in terms of features, objects, and scenes. The success of image analysis depends on reliability of segmentation, but an accurate partitioning of an image is generally a very challenging problem.

Segmentation techniques are either contextual or non-contextual. The non-contextual techniques take no account of spatial relationships between features in an image and group pixels together on the basis of some global attributes, viz. grey level or colour. On the other hand, the contextual techniques group together pixels with similar grey levels and close spatial locations. A plethora of classical techniques are available in the literature for faithful segmentation of images. However, each of these techniques suffers from several drawbacks affected by the inherent uncertainties in images. The soft computing paradigm is very capable of handling the uncertainties prevalent in the image segmentation problem faced by the computer vision research community. However, the existing soft computing methodologies applied for the segmentation of multilevel and colour images are jeopardised in several respects. The most notable among them is the selection of the class levels/transition levels of the different segments/classes. Researches in this direction are aimed at finding suitable solutions to these problems.

An effort has been made in this book to remove the aforementioned short-comings of these soft computing methodologies involving neural network architectures, with special reference to the neighbourhood topology-based multilayer self-organising neural network (MLSONN) architecture through an optimization of the class levels/transition levels of the segments/classes. The book takes recourse to seven well-written chapters targeted in this direction.

Chapter 1 introduces to the audience the basics of image segmentation with reference to the classical approaches, the soft computing counterparts and the hybrid approaches. The chapter also discusses the different soft computing paradigms in brief like neural network, fuzzy sets and fuzzy logic, genetic algorithms and the classical differential evolutionary algorithm. Then it throws light on the role of single- and multi-objective optimization on image segmentation.

A brief review of the recent trends in image segmentation is illustrated in Chap. 2. The chapter starts with a discussion of the classical approaches of image segmentation. As a sequel, the chapter elaborates the different soft computing paradigm based image segmentation approaches in detail.

As earlier stated the main objective of the book is to remove the limitation of the existing soft computing paradigm as far as image segmentation is concerned. The initial steps in this direction are centred on the standard multilevel sigmoidal (MUSIG) activation function of the MLSONN architecture. This results in an induction of the data heterogeneity in the area of clustering/segmentation with the help of the conventional self-organising neural network. The MUSIG activation function uses equal and fixed class responses, assuming the homogeneity of image information content. Chapter 3 introduces a genetic algorithm based optimised MUSIG (OptiMUSIG) activation function, which enables the network architecture to incorporate the underneath image content for the segmentation of the multilevel greyscale images.

Suitable extensions to the OptiMUSIG activation function with the help of the MLSONN architecture for the purpose of segmentation of true colour images have been proposed by resorting to a parallel version of the activation function in Chap. 4. The genetic algorithm based parallel version of optimised MUSIG (ParaOptiMUSIG) activation function is generated with the optimised class boundaries for the colour components and is able to segment colour images efficiently with the help of the parallel self-organising neural network (PSONN) architecture.

Thirdly, in order to overcome the flaws of single objective based optimization procedures multi-objective based optimization procedures have been invested to solve the problem of image segmentation. A multi-objective based OptiMUSIG activation function has been presented in Chap. 5 to segment the multilevel grey-scale images. This refining procedure reduces the possibility of the non-effectiveness of a particular solution in the field of other objective functions. Not only restricted in this, a NSGA II based OptiMUSIG activation function is also presented to segment the multilevel gray scale images.

In attempt to put forward the aforementioned approaches together, a multi-objective genetic algorithm based ParaOptiMUSIG activation function, which obviates the shortcomings of the single objective based ParaOptiMUSIG activation function, is proposed in Chap. 6 to segment colour images. Similar to the NSGA II based OptiMUSIG activation function, a NSGA II based ParaOptiMUSIG activation function is also presented for the segmentation of true colour images.

A segmentation procedure with a predefined number of classes cannot assure good results. Good segmented output may be derived by increasing or decreasing the number of classes if the exact number of classes in that test image/dataset is unknown. This is when no *a priori* knowledge regarding the information distribution, the number of classes and the information about the class responses are supplied at the preliminary stage. In this direction a genetic algorithm based clustering algorithm is presented in Chap. 7 to perform automatic clustering/segmentation. The effectiveness of a cluster/segment is validated by a proposed fuzzy intercluster hostility index. The proposed segmentation process starts from a large number of classes and finds out the exact number of classes in the test image/dataset.

The experimental findings of each of the chapter reveal that the hybrid soft computing paradigms resorted to yield superior performance as compared to the soft computing counterpart devoid of any hybridization. This substantiates the fact that proper hybridization of the soft computing tools and techniques always leads to more effective and robust solutions since the constituent soft computing elements in the hybrid system always complement each other.

The authors have tried to bring together some notable contributions in the field of hybrid soft computing paradigm for the application of multilevel image and data segmentation. The authors feel that these contributions will open up research interests among the computer vision fraternity to evolve more robust, time-efficient and fail-safe hybrid intelligent systems. The authors believe that this book will serve the graduate students and researchers in computer science, electronics communication engineering, electrical engineering, and information technology as a reference book and as an advanced textbook for some parts of the curriculum. Last but not the least, the authors would like to take this opportunity to extend their heartfelt thanks to the editors of the Springer book series Computational Intelligence Methods and Applications, and to Mr. Ronan Nugent, Senior Editor, Springer-Verlag for his constructive support during the tenure of the book project.

Cooch Behar, India   Sourav De
Kolkata, India   Siddhartha Bhattacharyya
Howrah, India   Susanta Chakraborty
Santiniketan, India   Paramartha Dutta
August 2016

Contents

Chapter 1
Introduction

1.1 Introduction

Human civilisation has been witnessing different phases of discoveries in course of its development. As a result, society progressed and transformed better. With the knowledge content of what has been discovered yesterday, sets in the possibility and scope of today's discovery. Getting inspiration from what is discovered today, human beings dream for new finding of tomorrow. Naturally, the development of human civilisation is a process in continuum. In a crucial phase of human civilisation, discovery of number system and introduction to counting were formidable findings. This is more because its consequence has been found far reaching. What we observe today as the ingredients of Computational Science, they have got the origin there only. What has changed over time is the mechanism and paradigm of computation and not the foundation. Capability to compute, apart from others happens to be an edge enjoyed by human beings over other living creatures. Subsequent discovery of devices with computational capability of huge quanta, mostly complex, of computation has supplemented and thereby offered relief to humankind. Such computational devices have undergone change over time with finding of more and more sophistication and improvisation. Civilisation witnessed more and more powerful computing devices over time—commonly referred to as computers of generations one after another. If the spirit of evolution of various generations of computers is traced, it is the endeavour to achieve alternative to what is typically available in a human being. It is needless to mention that such a substitute would typically be artificial in respect of its behaviour. The question then started peeping human civilisation is in case it would be possible to arrange computation in an artificial device then would it be too ambitious to think of realising intelligence even in an artificial manner. Subsequently, the challenge of imparting "intelligence" in a computing device caught the imagination of research community. The fruit of their effort gave birth to "Artificial Intelligence". With the tireless endeavour put in by the relevant research community the field of Artificial Intelligence has now been able to attain a

© Springer International Publishing AG 2016
S. De et al., *Hybrid Soft Computing for Multilevel Image and Data
Segmentation*, Computational Intelligence Methods and Applications,
DOI 10.1007/978-3-319-47524-0_1

vast shape. It appears quite prudent to understand as to what ingredients were there which could convert a machine, none other than a device, an intelligent entity as an effect of embedding intelligence artificially. It is obvious that as a result of this, a machine expectedly started behaving intelligently, to a limited extent albeit. It goes without saying that such attempt could not have been successful but for the formal mechanism of learning a machine. Such an effort culminated an area of research identified as "Machine Learning". In other words, "Machine Intelligence" is the natural consequence of "Machine Learning".

By this time, it has become clear to us that these three, viz. "Artificial Intelligence", "Machine Learning" and "Machine Intelligence" took shape one by one, in the said order chronologically. There has been serious research effort in these areas with formidable reporting. The real-life applications of these technical ingredients are found all pervading. In fact, in the present phase of human civilisation, with extensive automation around, we are unable to think computing systems devoid of pinch of "intelligence" in it. This is enough to indicate that how we have become dependent on "intelligent" systems in our day-to-day life. In fact, it is unimaginable to think of reverting this course of development and transition of the human civilisation from today's intelligent into tomorrow's intelligent one. It is better to orient our thinking otherwise, in the positive direction. It is judicious to embrace intelligent systems more and more, putting effort to have more intelligent systems around us in days to come and derive the fruit of research findings out of it.

The question as to what are the avenues to achieve intelligence in a machine comes next. Broadly speaking, there are primarily a few issues associated. In real-life scenario, we have to come across issues and problems most of which suffer from non-exactness. Classical mathematics is quite rich to provide solution methods capable of addressing most of the problems of exact nature. In fact, these methodologies dealing with exact problems appear, in the main, robust on the whole. In ideal situation, this is understandable. Unfortunately, these techniques fall short of offering comprehensive solution to most of the real-life problems, because of the non-exactness inherently associated to such problems. The need of next generation of computing paradigm, competent to accommodate such non-exact problems in its purview became essential. Zadeh [1] did introduce the notion of fuzzy mathematics and fuzzy logic as the first step towards dealing with non-exact scenarios effectively [2–6]. Subsequent to this, there has been reporting of so many important articles and it is still continuing. This is because the inherent source of non-exactness may be manifold —due to imprecision, vagueness, uncertainty so on and so forth with very blurred line of demarcation prevailing between them. Characterising a real-life problem, in respect of its specific nature of non-exactness appears to be a difficult task, if not impossible, most of the time. Naturally, coming out with a solution and that also robust in nature might appear difficult. Given a problem, what seems to be a good solution on different instances today, may not be found working well with other instances tomorrow. As a result, there is every scope of improvement of solution methodology corresponding to a problem, leading thereby the event of saturation of solution methodologies to any such problem remote. Accordingly, the spurt of research in this domain is formidably visible even today. Methods encompassing higher order fuzzy mathematics, Rough set, Vague set are reported as generation subsequent to fuzzy set. What appears to be the real challenge is the way to make the machine aware of such sources of non-exactness pertaining to a problem, followed by its effective solution strategy, to expect the machine behave intelligently in such contexts. As has already been indicated, conversion of a computing device into an intelligent entity needs machine learning. Question is how to learn a machine. Artificial Neural Network (ANN), the artificial analogue of human neural system has been observed to have worked exceedingly well in some contexts of machine computation. The literature reported that ANN has been able to offer a very robust computation platform towards achieving the task of learning a machine. Most of the relevant research initiative involves the issue of classification using ANN-based techniques. Once learnt, especially by appropriate ANN technique, the machine has been found to behave intelligently by classifying and thereafter recognising objects of interest quite nicely with high degree of accuracy. Two aspects viz. model and architecture become important while designing an ANN earmarked for a suitable task of recognition. The better are these two achieved, more appropriate expected is the recognition. Based on the technical requirement of a classification problem, there are several ANN models and/or architectures reported. Important among them are Perceptron, Hopfield network, Kohonen's Self-Organising Map (SOM), Radial Basis Function (RBF), Self Organising Neural Network (SONN) to name a few. Of course, there are others also. Support Vector Machine (SVM) has been reported and found to function extremely convincingly as classifier in typical situations. However, the journey is not exhausted and even today there are reporting of efficient ANN-based techniques.

Unfortunately, real-life situations do not remain content with dealing with problems of non-exact nature and classification only. Other challenging aspects comprise search and optimisation, which are beyond the purview of the computational platforms we discussed so far. Evolutionary Computation has the jurisdiction of dealing with search and optimisation issues. Classical approaches, in spite of possessing formidable potential in solving wide spectrum of such problems, do suffer from the shortcoming of being versatile. In fact, problems having associated huge and/or complex search space often fall short of rendering justice to the expectation of offering effective solutions in efficient manner. For example, typically problems belonging to NP-complete or NP-hard classes may not aspire for comprehensive solutions within reasonable amount of time resource. We know that theoretically finding exact algorithms to such problems in polynomial time is not possible until and unless, of course, it is possible to establish that Class P is identical to Class NP. In the absence of any polynomial time algorithm for such a problem, the reasonable alternative that we have to remain satisfied with is either a heuristic technique or an approximation algorithm. The underlying rationale in such context is that we have to get hold of a solution, even though not the best one (suboptimal solution), in a reasonable time limit. The paradigm of evolutionary computation handles with approaches and techniques which primarily work in a stochastic manner based on some effective (may be context specific) heuristic to arrive at a solution preferably optimal, otherwise suboptimal. Use of the term "evolutionary" is justified because an algorithm wedded to the philosophy of evolutionary computation, tries to arrive at a solution to a problem in evolutionary manner, i.e. the underlying search process moves towards a solution in the underlying solution space, from generation to generation giving rise to evolution. Genetic Algorithm (GA), propounded by Holland was the first evolutionary approach reported in the literature. The fundamental spirit behind the operability of GA, as per him was the artificial analogue of the natural genetic process. The Darwinian principle of "survival of the fittest" principle, prevalent in the Nature is reflected in the reproduction operation through the process of selection in the functioning of GA. In GA, there are operators such as crossover and mutation also operative. Whereas the crossover operator is the mimic of recombination, functioning as an indispensable component of natural genetics, the genetic mutation in the Nature has its artificial counterpart in the form of mutation operator in GA. With these three GA operators viz., selection, crossover and mutation, operative in unison, offer the phenomenon of evolution of generations of the GA, moving towards a solution in the search space corresponding to an optimisation problem. Various optimisation problems with GA based solutions are available in literature. There are subsequent variants of GA which are found to have engrossed the literature heavily. There is also some reporting of a number of articles exploring the theoretical aspects of GA and variants, in terms of their efficiency, performance, etc. Gradually, the shortcomings of GA started becoming visible. As a result of serious research, people could arrive at the inference that the spirit of evolutionary process pursued by GA being a technique inspired by Nature, has substantial merit even though, has its limitations also, particularly due to its slow and time-consuming performance. Gradually, came into reporting other evolutionary approaches such as Simulated Annealing (SA), Particle Swarm Optimisation (PSO), Ant Colony Optimisation (ACO) etc. with each indicating respective strength over GA counterpart, in respect of time. It is needless to mention that all these methods are characteristically evolutionary and as such are part and parcel of Evolutionary Computation platform. With the growing demand of solution methodology with multiple objectives in simultaneity, commonly known as multi-objective optimisation, different evolutionary approaches witnessed the multi-objective variants also. They include Multi-objective GA (MOGA), Multi-objective Simulated Annealing (MOSA), Multi-objective Particle Swarm Optimisation (MOPSO), etc. There are many reporting on Multi-objective Evolutionary Optimisation (MEO) algorithms applied on different real-life problems offering encouraging results.

With the advantages of the Soft Computing methodologies encompassing Fuzzy mathematics, ANN, Evolutionary Computation used extensively in various real-life applications, people also started finding the shortfalls of them side by side. Gradually, it was felt that in spite of enormous effectiveness of these Soft Computing techniques at individual level, there are various real-life problems falling short of getting solved by exclusive use of an individual Soft Computing technique. Researchers were encouraged to explore avenues to overcome such shortcomings. It is in this backdrop, the era of Hybrid Soft techniques evolved. The question of exploiting the strength of one Soft Computing technique for supplementing the drawback of another started haunting the research community giving rise to hybridisation. In other words, it was felt that collaboration rather than exclusion of different Soft Computing techniques properly could be effective specifically for those real-life problems where individual technique have been found to have failed to render service. In fact, we started witnessing phenomenal growth in research initiative towards Hybridisation. The justification in favour of hybridisation is no longer entertained, with everybody accepting its underlying potential in several contexts of real life. Today the research community is more inclined to explore as to which form of hybridisation is effective and why. Hybrid techniques in the form of Neuro-Fuzzy, Fuzzy-Genetic, Neuro-Genetic and many more got their formidable presence in the literature. The inherent potential of hybridisation is definitely not yet fully explored. A lot of initiative is still to arrive. With the active involvement of relevant research community, we are sure to come across in days to come strong and effective presence of Hybrid techniques applied extensively on various real-life applications. The present book is on a specific application of image segmentation. In course of our discussion ahead, we shall try to explore as to how intelligent techniques may be made effective for this purpose.

1.2   Different Approaches Used for Image Segmentation

Based on the ways, the problem of image segmentation has been attempted solution, comprise-

- Classical approaches
- Soft Computing approaches
- Hybrid approaches

We shall try to provide a brief description of each approach individually.

1.2.1   Classical Approaches

We have some interesting reporting based on classical approach indicated in this context. In [7], authors provide a novel image segmentation technique particularly for image querying. They propose a novel representation, Blobworld, obtained through fully automatic clustering process, is coherent in the feature space comprising colour, texture and position. Authors in [8], propose a new multiphase level set framework for image segmentation using the Mumford and Shah model, for piecewise constant and piecewise smooth optimal approximations. A predicate for measuring the evidence for a boundary between two regions using a graph-based representation of the image is defined in [9], on the basis of which, an efficient segmentation algorithm is developed. Authors show that although this algorithm makes greedy decisions it produces segmentations that satisfy global properties. Another very interesting technique due to Kitney et al. is reported in [10]. Authors propose an efficient threshold selection method that is capable of working automatically from the image content [11]. Chosen thresholds are used for segmentation. Article [12] reports the fusion of colour information derived out of applying K-means based clustering for image segmentation. In [13], authors share very important information regarding medical image segmentation. It is informative because of its comprehensive reporting on automated approaches in this context. In [14], authors propose a generic image segmentation approach that is capable of addressing both the challenges of contour detection and segmentation at one go. Here, contour detector is converted into a hierarchical region tree, by which the authors reduce the problem of image segmentation to that of contour detection. In [15], authors propose a hierarchical graph-based image segmentation relying on a criterion popularised by Felzenszwalb and Huttenlocher. They have justified the effectiveness of their approach on different benchmark image data. In addition, an extensive and informative survey may be had in [16].

1.2.2   Soft Computing Approaches

In the present section, we shall try to share some brief information about some important literature reporting. A very effective and informative review article in years back may be had in [17]. In [18], the authors propose an edge-based segmentation computationally efficient algorithm built on a new type of active contour that does not introduce unwanted smoothing on the retrieved contours. The contours are always returned as closed chains of points, resulting in a very useful base for subsequent shape representation techniques. Practically in the contemporary time, we find another important contribution in [19]. Another very important survey article at that time is available in [20]. Authors report object tracking methodology in video environment in [21]. We come across a novel shape based approach in [22]. Another information paper of comparatively recent time is [23]. Boundary code is found to have been used for edge detection towards achieving image segmentation in [24]. There is many other reporting also. However, here we did indicate only a few important out of them. In subsequent detailed description we shall come across some more.

1.2.3   Hybrid Approaches

The method in [25] is based on the computation of partial derivatives obtained by a selective local biquadratic surface fit. Subsequently, the Gaussian and mean curvatures are computed on the basis of which initial region-based segmentation is obtained in the form of a curvature sign map. Two additional initial edge-based segmentations are also computed from the partial derivatives and depth values. The three image maps are then combined to arrive at the final segmentation. A wonderful review article may be had in [26] due to Bezdek et. al. This review is confined to addressing segmentation of MR images only. However, it is very informative from medical research point of view. Authors in [27], propose a two-stage procedure with an edge-preserving statistical noise reduction approach as preprocessing stage for achieving an accurate estimate of the image gradient followed by partitioning the image into primitive regions by applying the watershed transform on the image gradient magnitude. This initial segmentation is fed as input to a computationally efficient hierarchical region-merging process to arrive at the final segmentation. Piccardi offered an effective review [28] for identification of background in a video environment. The challenge of such problem is its segmentation and extraction procedure in dynamic environment. Tan et al. offer a novel approach to achieve the task of colour image segmentation [29]. Their hybrid technique based on Fuzzy C-means has been found to work exceedingly well for such segmentation with thresholding being done from raw histogram information. A new perspective altogether has been reported in [30]. The authors here introduce a novel taxonomy based on the amount of shape knowledge being incorporated in the segmentation process. Accordingly, their argument is that all global shape prior segmentation methods are identical to image registration methods and that such methods cannot be characterised as either image segmentation or registration method separately. They propose a new class of methods that are able solve both segmentation and registration tasks in simultaneity, called *regmentation*. In [31], authors propose a novel hybrid paradigm, with four techniques clubbed within, based on a new form of interaction called live markers, where optimum boundary tracking segments are turned into internal and external markers for region-based delineation to effectively extract the object. They provide four techniques within this paradigm: (1) LiveMarkers; (2) RiverCut; (3) LiveCut; and (4) RiverMarkers. There is reporting of a very recent edited book by Bhattacharyya et. al. [32], containing a number of important collections in the form of book chapters such as [33–35]. Authors propose a novel activation function to achieve segmentation of colour MR images in [36]. Whereas [34] reports on informative contribution on human face recognition, [35] provides an important finding regarding a comprehensive study and analysis on identification of breast cancer in thermogram images. These articles offer insight into the very recent trend of the relevant field.

1.3 Soft Computing Techniques

1.3.1 Neural Network

Millions of neurons or very simple processing units are linked together in brain in parallel manner. These neurons are responsible for the human intelligence and recognition power. Neurons are connected with each other by a direct communication link, each with an associated weight. The neural networking paradigm is an intelligent reasoning and decision-making system that is based upon the biological neural networks in that the operational characteristics of neural networks more or less similar with the behaviour of human brain. Basically, neural networks [37–40] is a parallel and layered interconnected structure of a large number of artificial neurons that interact with the real world in the same manner with the biological systems [41]. Each and every individual neuron makes an elementary computational primitive. Kohonen [39, 42, 43] presented neural networks as *Artificial neural networks are massively parallel adaptive networks of simple non-linear computing elements called neurons which are intended to abstract and model some of the functionality of the human nervous system in an attempt to partially capture some of its computational strengths*. The interconnection topology of the neurons determines the corresponding flow of information in a neural network and the signals flow between neurons through the interconnection links. In a distinctive neural net, the transmitted signals are multiplied with the associated weight of each connection link. The weights constitute information being used by the net to solve a problem. An activation function plays a vital role to determine the output signal when it is applied on the sum of weighted input signal of each neuron. A neural net is categorised on the basis of three criteria [44], first, its connection patterns between the neurons, second, its method of determination of weights associated to the connections, and finally, its activation function. The structures of the neural networks are different from one another on the basis of the topology of the underlying interconnections as well as on the target problem they are going to solve. As the human brain has the learning capability from the environment and the remembering capability, these neural networks also have the learning capability like the human brain by the form of training by the test information. These networks are trained with the different aspects about the problem, like the input-output data distribution and relationships. After that, that network can be employed to solve similar type of problems. Different neural network architectures are employed on the basis of the nature of the problem and depending on the type of the learning procedure adopted.

Generally, neural network is a layered structure of neurons and consists of the following eight components [37, 45, 46]:

- *Neurons*: From the previous discussion it has been noted that the neurons are the basic computational elements of a neural network [45–47] and there are three types of neurons, viz. input, hidden and output. The outside world information are fed into the input layer of the neural network and the hidden layer neurons operate on and process the data fed by the input neurons. The hidden layer resides in between the input and output layers. After processing the data by the hidden layer neurons, the processed data are transferred to the output layer neurons and the output layer neurons generate the output result finally [48, 49].
- *Activation state vector*: As expressed earlier, the excitation/activation levels of the nerve cells or neurons play a vital role during the transmission of information through the different neurons of a human nervous system. The same scenario employs to the artificial neural networking paradigm as well [48]. The activation levels of the neurons of the network are determined by the neuron activation state

vector. The activation vector ($X \in R^n$) is denoted as [46, 48] $X = \begin{bmatrix} x_1 \\ x_2 \\ . \\ . \\ x_n \end{bmatrix}$, if there

are n number of neurons $x_i, i = 1, 2, \ldots, n$.

- *Activation function*: The transfer characteristics of the individual neurons in a neural network are presented by this function and this function is applied to determine the activation behaviour of the neurons to incident inputs [46, 48]. Generally, the functions are applied to respond to the range of incident input signals and a suitable learning algorithm is used to transform them. These activation functions may vary from neuron to neuron within the network. Most of the popular neural network architectures are field-homogeneous [48, 49], i.e. all the neurons within a layer are characterised by the same activation function. The binary threshold, sigmoid, linear threshold, probabilistic activation functions, etc. are some of the commonly used standard activation functions. The interested readers can get the description of those standard activation functions in the articles [45, 46, 48].
- *Interconnection Topology*: This defines the mode of interconnections of the neurons in different layers of a neural network architecture. It is similar to the axon interconnections in biological nervous system but not in complex manner. These interconnections behave as the storage junctions/memory of the artificial neural network. These interconnections are also characterised by a weight value like the synapses of biological neurons which determine their excitation/activation levels [46, 48, 50].
- *Neuron firing behaviour*: The processing capabilities of the neurons in a given layer, the characteristic activation function applied therein are determined by this component [46, 48]. The constituent neurons combine the incoming input signals by multiplying the input vector and the neuron fan-in interconnection strength (weight) vector and impress the embedded activation function.
- *Learning algorithm*: Neural networks have the capability to learn from the external example. The way of training a neural network with a training set of input–output pairs of data relationships is noted as the learning rule [46, 48]. The aim of this type of algorithms is to improve the performance. The objective of the algorithm is to indemnify the network system error by modifying the interconnection weights and enable the network to arrive at the desired end result. The type of learning algorithm can be classified in three main categories, viz., supervised learning, unsupervised learning and self-supervised learning.

  - *Supervised learning*: Basically, this type of learning is done on the basis of the direct comparison of the network output with known correct or desired answer and the interconnection weights are adjusted gradually.
  - *Unsupervised learning*: Without knowing the whereabouts of the relationship of input-output, the network updates in unsupervised learning by a complex competitive–cooperative process where the individual neurons compete and

cooperate with each other to update their interconnection weights during the
process of self-organisation [49].

- *Self-supervised learning*: Self-supervised learning is a special kind of supervised learning and it is also known as reinforcement learning. In this learning
process, the network tries to learn the input-output mapping through trial and
error process and seeks to maximise the performance. The network has an idea
whether the output is correct or not, but does not have any idea of correct output [41].

Neural networks are employed to perform different type of tasks, like, clustering,
pattern classification, optimisation, prediction or forecasting, function approximation, content retrieval, to name a few [41]. Like the weighted directed graph, artificial
neurons are considered as nodes and the connection between neuron inputs and outputs are noticed as the directed edges with weight. Commonly, the neural network
architectures are categorised in the following two classes on the basis of the interconnection topologies.

- *Feedforward neural network*: This type of neural network does not have any loops.
  The single layer perceptron, multilayer perceptron [37, 46, 51], radial basis function networks (RBFN) [52], support vector machines (SVM) [53, 54], etc. are
  few examples of this type of neural network. These neural networks apply different kind of learning algorithms, like, the perceptron learning, least mean squares,
  backpropagation learning, reinforcement learning, support vector learning and
  their variants [55, 56].
- *Feedback neural network*: This type of network has the loops due to feedback connection. The Hopfield network [57], brain-in-a-state-box (BSB) mode [58], Boltzmann machine [59], bi-directional associative memories (BAM) [60], adaptive resonance theory (ART) [61], Kohonen's self-organising feature maps (SOFM) [42]
  are some of the typical examples of this type of network.

Artificial neural networks are resistant to noise, able to recognise partially degraded
images, able to classify overlapped pattern classes, tolerant to distorted images or
patterns and have the potentiality for parallel processing.

1.3.2   Fuzzy Sets and Fuzzy Logic

Most of the information available in the real world are very much complex and
the complexity comes from uncertainty in the form of ambiguity [62]. In our daily
life, information are transmitted in different forms, like, data, images, video content,
speech signals or in any other electronic forms. These consumed real-world data
comprise a varied amount of ambiguity and imprecision, which cannot always be
measured in practice. In those cases, a classical computing system is not acquainted
with the associated uncertainty and imprecision to the principles of finiteness of
observations and quantifying propositions applied. The fuzzy sets and fuzzy logic,

another paradigm of the soft computing, render a logical framework for the description of the varied amount of ambiguity, uncertainty and imprecision presented in real-world data under consideration. Human beings are capable to handle the complex and ambiguous problems subconsciously from the early days of life. They can achieve an approximate decision out of a reasoning framework based on experience and linguistic expressions; a capability that was tried to incorporate in the computer system. In fact, the *fuzzy set* approach equips with the linguistic modes of reasoning that are natural to human beings [45]. The fuzzy set theory, introduced by Professor Lotfi Zadeh [62–68]. The vagueness and ambiguity inherent in everyday life data are efficiently handled by Fuzzy set theory. This theory contradicts the concept of crisp sets, where information is more often expressed in quantifying propositions. As per professor Zadeh, the foundation of fuzzy sets and fuzzy logic is fully dependent on the intelligence in human reasoning. He was incited by the fact that human beings more often communicate via natural language terms or linguistic expressions, which cannot be always quantified by numeric values [2, 63, 67–69]. To include precise mathematical analysis, this theory renders an approximate and yet effective means for describing the characteristics of a system that is too complex or ill-defined [41, 70, 71]. Fuzzy logic is a superset of conventional (Boolean) logic that has been extended to handle the concept of partial truth, i.e. truth values between "completely true" and "completely false". This fizzy approach assumed that the human thinking cannot be represented only by numbers but can be guessed to tables of fuzzy sets, or, in other sense, classes of objects in which the transition from membership to nonmembership is gradual rather than abrupt. The essence of fuzzy logic lies from the fact that most of the logic behind human reasoning is not the two-valued or even multi-valued logic but logic with fuzzy truths, fuzzy connectives and fuzzy rules of inference [41]. The basic concept behind the *fuzzy sets* [2, 62, 63] is that any observation which presents varied amount of uncertainties, exists with a varied degree of containment in the universe of discourse. This degree of containment is pertained to as the membership value of the observation. A fuzzy set is a mapping from an input universe of discourse into the interval [0, 1] that depicts the membership of the input variable. This mapping is referred to as fuzzification. A crisp information is represented into the fuzzy way by a tool, named as fuzzification. The defuzzification is a reverse mechanism to revert to the valued crisp logic from the fuzzy logic.

Based on the theory of fuzzy sets, fuzzy logic is a multi-valued logic [65], which incorporates all possible outcomes corresponds to an observation like classical logic. It handles the inaccurate modes of reasoning and thought processes with linguistic variables. Decision support and expert systems with powerful reasoning capabilities are provided by the fuzzy techniques in the form of approximate reasoning. The human beings are capable to make rational decisions in an environment of uncertainty and imprecision. Based on the previous knowledge that is not accurate, not complete or not totally satisfactory, an approximate answer can be given to a question. Everything, including truth and false, is a matter of degree in fuzzy logic. Any logical system can be fuzzified [45]. A theory of approximation reasoning was developed by Zadeh [63] using fuzzy set theory. Using this approximation reasoning, any kind of reason is presented neither very exact nor very inexact [41]. Human reasoning and thinking process can be modelled by this theory with linguistic variables and they can deal with both hard and soft data as well as various type of uncertainties [70]. According to fuzzy logic, knowledge is represented as a collection of elastic and a fuzzy inference is regarded as a process of propagation of elastic constraints.

Fuzzy rules is collection of conditional statements of fuzzy logic and they create the basis of the linguistic reasoning framework, which embodies representation of shallow knowledge. In natural language or in linguistic, the fundamental atomic terms are often altered with adjectives or adverbs and it is not an easy task to list them all [62]. These modifiers are denoted as "linguistic hedges". These linguistic hedges have the effect of modifying the membership function of a basic atom. The common way to represent a fuzzy rule [62], which is similar to natural language expressions, can be noted as

IF premise (antecedent) THEN conclusion (consequent)

Generally, this type of expression is referred to as the IF-THEN *rule based* form. It typically expresses an inference such that if a fact (premise, hypothesis or antecedent) is known, then another fact (conclusion or consequent) can be derived. This type of knowledge representation, represented as shallow knowledge [62], is quite suitable in the context of linguistics because it expresses human empirical and heuristic knowledge in the language of communication.

A generic fuzzy system develops from fuzzy set-theoretic concepts and fuzzy rule guided logical reasoning. It consists of the following modules.

- A *fuzzification* interface [45] that fuzzifies the numeric crisp inputs by assigning grades of membership using fuzzy sets defined for the input variable.
- A *fuzzy rule base/knowledge base* [45] that consists of a data-derived or heuristic rule base. The clustering techniques or neural networks generates the data-derived rule base using sensor databases. On the other hand, the human experts generates the heuristic rule base through some intuitive mechanisms.
- A *fuzzy inference engine* [45] that generalises fuzzy outputs by utilising the fuzzy implications and the rules of inference of fuzzy logic.
- A *defuzzification interface* [45] that concedes a non-fuzzy crisp control action from an inferred fuzzy control action.

The applications of fuzzy systems find in a wide variety of disciplines which include control, signal processing, function approximation, time series prediction, etc.

1.3.3 Fuzzy Set Theory

A fuzzy set [62, 63] is a collection of elements, $A = \{x_1, x_2, x_3, \ldots, x_n\}$ charac- terised by a membership function, $\mu_A(x)$. An element, x, has a stronger containment in a fuzzy set if its membership value is close to unity and a weaker containment therein if its membership value is close to zero. A fuzzy set A, comprising elements $x_i, i = 1, 2, 3, \ldots n$ with membership $\mu_A(x_i)$, is mathematically expressed as [62, 63]

$$A = \sum_i \frac{\mu_A(x_i)}{x_i}, i = 1, 2, 3 \ldots n \tag{1.1}$$

where $\sum_i$ represents a collection of elements.

The set of all those elements whose membership values are greater than 0 is referred to as the support $S_A \in [0, 1]$ of a fuzzy set A. S_A can be denoted as [62, 63]

$$S_A = \{ \sum_i^n \frac{\mu_A(x_i)}{x_i} : x_i \in X \text{ and } \mu_A(x_i) > 0 \} \tag{1.2}$$

The maximum membership value of all the elements in a fuzzy set A is referred to as the height (hgt_A) of the fuzzy set. A fuzzy set is a normal or subnormal fuzzy set depending on whether hgt_A is equal to 1 or is less than 1.

The normalised version of the subnormal fuzzy subset (A_s) can be expressed by [72]

$$\text{Norm}_{A_s(x)} = \frac{A_s(x)}{hgt_{A_s}} \tag{1.3}$$

The normalisation operator for a subnormal fuzzy subset A_s with support, $S_{A_s} \in [L, U]$ is expressed as [72]

$$\text{Norm}_{A_s(x)} = \frac{A_s(x) - L}{U - L} \tag{1.4}$$

The corresponding denormalisation can be attained by [72]

$$\text{Denorm}_{A_s(x)} = L + (U - L) \times \text{Norm}_{A_s(x)} \tag{1.5}$$

The normalised linear index of fuzziness [3] of a fuzzy set A having n supporting points, is a measure of the fuzziness of A. It is given by the distance between the fuzzy set A and its nearest ordinal set $\underline{A}$. It is given by [3]

$$v_l(A) = \frac{2}{n} \sum_{i=1}^n |\mu_A(x_i) - \mu_{\underline{A}}(x_i)| \tag{1.6}$$

i.e.

$$v_l(A) = \frac{2}{n} \sum_{i=1}^n [\min\{\mu_A(x_i), (1 - \mu_A(x_i))\}] \tag{1.7}$$

The subnormal linear index of fuzziness (v_{l_s}) [72] for a subnormal fuzzy subset A_s with support $S_{A_s} \in [L, U]$, is given by

$$v_{l_s} = \frac{2}{n} \sum_{i=1}^{n} [\min\{(\mu_A(x_i) - L), (U - \mu_A(x_i))\}] \qquad (1.8)$$

1.3.4 Genetic Algorithms

Genetic algorithms (GAs) [73, 74] are efficient, adaptive and robust stochastic search and optimisation techniques for complex non-linear models where fixing of the global optimum is a difficult task. Basically, GAs are applied to solve the objective or fitness function in large, complex and multimodal search spaces. Instead of getting trapped into local optimum solutions, the GAs attempt to get the global optimum solutions of a problem. The natural genetics and evolutionary processes are the working principle to produce several solutions to a given search problem. To derive the global solution, GAs start with a set of probable solutions and they are processed in parallel. The probable solutions are encoded in a fixed length based bit string, known as *chromosome*. The chromosomes are used to carry the complex information of the search space. Each chromosome is associated with a *fitness* value. An *objective/fitness function* associated with each string, provides a mapping from the chromosomal space to the solution space. GAs are suitable for solving optimisation problems. Potentially better solutions can be derived by applying various biologically inspired operators like *selection*, *crossover* and *mutation* on the chromosomes.

1.3.4.1   Basic Precepts and Characteristics

Genetic algorithms follow biological principles to solve complex optimisation problems. GA starts with a set of possible solutions or chromosomes as a *population* or *gene pool*. These chromosomes or the individual solutions are generated randomly. These chromosomes are processed by some biologically inspired genetic operators to generate a new set of solutions from the previous one. This is motivated by the theory of evolution, the better suited individuals in a population may survive and generate offspring in the environment. Therefore, the superior genetic information are channelised to the next generation to the particular problem. GAs are defined by the following essential components:

- The *chromosomes*, formed in a string-like structure, used to generate in such a way that it hold information about the potential solutions of a search space.
- A set of probable solutions or *chromosomes* are known as population or *gene pools*.
- *Fitness function*, a evaluation mechanism of every individual in the solution domain.

- Mechanism for selection of better solutions (*selection*).
- *crossover* operator.
- *mutation* operator.

The different steps of a GA is as follows:

1. Random population initialisation
2. Compute the fitness of population using relevant fitness function
3. Repeat (until a certain is not satisfied or for a predefined number of iteration)

   a. Select parents from population, on the basis of fitness values
   b. Execute crossover and mutation to create new population
   c. Compute the fitness of new population
   d. Go to Step 2

The components are illustrated briefly in the following subsections.

1.3.4.2 Chromosome Encoding and Population Generation

To solve a problem, GAs use a population of candidate solutions like the traditional search methods. Different types of encoding representation are applied on a chromosome to replicate information about the solution it represents. Depending upon the problem, the size of the chromosomes may be fixed or variable and they can vary from generation to generation. The size of population also affects the performance and scalability of genetic algorithm. The domain knowledge or any other information can be easily induced to the candidate solutions of the initial population. The chromosomes can be created using binary data, real-valued data, character data, etc. For example, we want to handle a problem and the value of the genes in a chromosome may vary from *0* to *10*. Suppose, four bits are needed to represent those values in binary and two genes can make a chromosome. So, the length of the chromosome is eight. To create a chromosome, a randomised function is applied and after that, 6 and 5 are generated. These decimal values are converted into binary to create a particular chromosome. So, a binary coded chromosome of length eight may look like

$$0 \ 1 \ 1 \ 0 \ 0 \ 1 \ 0 \ 1$$

1.3.4.3 Evaluation Technique

The fitness function determines the quality/suitability of a chromosome as a solution to solve a particular problem. In other words, the fitness function is characterised over a genetic representation that measures the quality of the represented solution. The selection of the fitness function is a crucial step and they are totally dependent on the problem to be solved. The fitness value f_i of the ith member denotes the quality of that candidate solution with respect to that problem. Subsequently, these fitness values are applied by GA for the guidance of the good solution evaluation.

1.3.4.4   Genetic Operators

Like the natural genetics, GAs also apply three genetic operators, viz. selection, crossover and mutation operators. The new potential child populations are generated after applying these operators on the population of *chromosomes*. The operators are illustrated in the next subsections.

1. *Selection*: In the selection/reproduction step, a tentative new population is generated after copying individual solutions from the initial population. This new population is known as *mating pool*. On the basis of the survival-of-the-fittest mechanism, the fitness value of the individual solutions are totally responsible for the selection of the number of copies for the next generation. A selection probability proportional to the fitness value is used to select the parent members for mating. The selection of a chromosome with higher fitness value has a higher probability to contribute one or more offsprings in the next generation. There are various selection methods viz., roulette wheel selection, stochastic universal selection, Boltzmann selection, rank selection, etc [73]. The main objective of this step is the improvement of the quality of the solutions over successive generations.

2. *Crossover*: The main intension of the crossover operation is to interchange information among the randomly selected parent *chromosomes* from the mating pool. The parts of the genetic information of two or more parental solutions are recombined to produce new, possibly better offspring and they are propagated to the next generation of the population. Parents with equal probability of crossover rate are selected for breeding. Various types of crossover operators are found in the literature [73]. The basic *single-point crossover*, uses a randomly selected crossover point. The two strings are divided into heads and tails at the crossover point. The tail pieces are swapped and rejoined with the head pieces to produce two new strings. For example, two chromosomes, say Ch_1 and Ch_2, are selected randomly from the mating pool and they behave as the parents. The crossover probability, p_c, is the predefined probability and used to take decision that crossover occurs between two parent chromosomes. For this, a random number $rn \in [0, 1]$ is generated. A crossover happens if $rn \leq p_c$ otherwise the parent chromosomes are copied for the next stage. The *crossover point, pt,* is selected randomly as an integer between 1 and $CL - 1$, where CL is the chromosome length. Now, the offsprings are produced by swapping the segments of the parent chromosomes from the ptth position to CL. Let two parent chromosomes and the crossover points be as presented below.

$$0\ 1\ 1\ 0\ 0\ 1|\ 0\ 1$$
$$1\ 0\ 0\ 1\ 1\ 0|\ 1\ 0$$

Then after crossover the offsprings will be the following:

$$0\ 1\ 1\ 0\ 0\ 1\ 1\ 0$$

$$1\ 0\ 0\ 1\ 1\ 0\ 0\ 1$$

Two point crossover, multiple point crossover, shuffle-exchange crossover and uniform crossover are some other crossover techniques [74]. The coding technique applied to represent the problem variables plays a vital role for the successful operation of GAs [75, 76]. The successful operation of GAs mainly depends on the coding–crossover interaction. The building block hypothesis indicates that GAs work by identifying good building blocks, and larger building blocks are generated by combining them finally [73, 77, 78]. The crossover operation cannot combine them together without good building blocks are coded tightly [79, 80]. Based on the tight or loose coding of problem variables, the problem is mostly known as the linkage problem [81].

3. *Mutation*: A random alteration in the genetic structure of a chromosome occurs in mutation with an objective that genetic diversity may be incorporated in the population. The process may not be able to attain the global optima if the optimal solution engages in a portion of the search space which is not symbolised in the population's genetic structure. In those cases, the possible solution for directing the population to the optimal section of the search space is mutation and it is done by randomly altering the information in a chromosome [45]. In case of binary chromosomes, the mutation operator flips some of the bits on the basis of the mutation probability p_μ. Using p_μ, a decision can be taken whether that candidate solution would undergo mutation or not. Suppose, position 2 and 5 are selected for mutation in an 8-bit chromosome and the bits at those points are complemented. The application of mutation on that chromosome is shown below.

$$1\ \mathbf{0}\ 0\ 1\ \mathbf{1}\ 0\ 1\ 0$$

$$1\ \mathbf{1}\ 0\ 1\ \mathbf{0}\ 0\ 1\ 0$$

For real coded genes, different techniques are considered for mutation. Basically, mutation provides insurance against the development of a uniform population incapable of further evolution.

1.3.4.5 GA Parameters

Several parameters in GAs like, the population size, probabilities of performing crossover (usually kept in the range 0.6 to 0.9) and mutation (usually kept below 0.1) and the termination criteria adjusted by the user. There are different types of replacement strategy, viz. the generational replacement strategy where the total population is replaced by the new population, or the steady state replacement policy where only the less fit individuals are replaced. These type of parameters and variation in different steps in GAs are totally problem-dependent and do not have any proper guidelines in different literatures. The repetition of the cycle of selection, crossover and mutation is done for a number of times till one of the following criteria matches [45]:

1. after a specified number of generations, average fitness of a population becomes more or less constant
2. Minimum a string in the population will attain the desired objective function.
3. number of generations is greater than some predefined threshold.

GAs find applications in the field of image processing, data clustering [82], path finding [83], project management, portfolio management [84] etc.

1.3.5 Classical Differential Evolution

Basically, differential evolution (DE) [85] is a well-known optimisation technique in the evolutionary computation family. This algorithm is a variant of genetic algorithms. It also applies biological operations, like crossover, mutation and selection operation on a population. It is different from classical GAs in two respects, e.g. the floating-point chromosomal representation instead of bit string representation of the same and arithmetic operation is applied instead of logical operation in mutation stage. For a certain number of generations, it has been attempted to maximise/minimise an objective criterion [78].

The candidate solutions or the individuals [85] are denoted as D-dimensional optimisation parameter vectors, i.e. $\overrightarrow{X}_{i,G} = \{x_{i,G}^1, \ldots, x_{i,G}^D\}, i = 1, \ldots, NP$ towards the global solution [85]. The number of members in a population and the number of generations are represented as NP and G, respectively. At the initial stage, the prescribed minimum ($\overrightarrow{X}_{min} = \{x_{min}^1, \ldots, x_{min}^D\}$) and maximum ($\overrightarrow{X}_{max} = \{x_{max}^1, \ldots, x_{max}^D\}$) parameter bounds are applied to generate the randomised individual solutions. It has been desired that these initial solutions can cover the entire search space [85, 86]. In classical DE, mutation operation is done before crossover and selection operation.

In the mutation operation, for each individual, $\overrightarrow{X}_{i,G}, i = 1, \ldots, NP$ at generation G, a mutation vector $\overrightarrow{A}_{i,G} = \{a_{i,G}^1, \ldots, a_{i,G}^D\}$ is determined by the following equation [85, 86]

$$\overrightarrow{A}_{i,G} = \overrightarrow{X}_{m,G} + AF.(\overrightarrow{X}_{n,G} - \overrightarrow{X}_{r,G}) \tag{1.9}$$

where, i, m, n, r are mutually different integers and randomly selected within the range $[1, NP]$. The scaling factor, AF, is applied as a positive mutation constant for the amplification of the difference vector [85, 86].

In the crossover, the trial vector $\overrightarrow{T}_{i,G} = \{t_{i,G}^1, \ldots, t_{i,G}^D\}$ is generated with the help of each pair of target vector, $\overrightarrow{X}_{i,G}$ and the same indexed mutant vector $\overrightarrow{A}_{i,G}$. The binomial/uniform crossover is defined as follows [85, 86]:

$$t_{i,G}^j = \begin{cases} a_{i,G}^j & \text{if } (j_rnd \le CE) \text{ or } (j = j_{rand}) \\ x_{i,G}^j & \text{otherwise} \end{cases} \tag{1.10}$$

where, $j = 1, 2, \ldots, D$ and j_{rand} is the randomly chosen integer in the range $[1, D]$.

The crossover rate [86], CE, is a user-defined constant in the range $[0, 1)$. It is applied to control the fraction of parameter values copied from the mutant vector.

All the trail vectors are evaluated on the basis of the objective function and after that, they will go for the selection operation. The selection operation is expressed as follows [85, 86]:

$$\overrightarrow{X}_{i,G+1} = \begin{cases} \overrightarrow{T}_{i,G} & \text{if } f(\overrightarrow{T}_{i,G}) \le f(\overrightarrow{X}_{i,G}) \\ \overrightarrow{X}_{i,G} & \text{otherwise} \end{cases} \tag{1.11}$$

where f denotes the objective function under consideration. Basically, the better individuals in the population are preserved by making a comparison between each individual of the trial vector with its parent vector and the better one is stored for the next generation.

Until some specified termination criteria are satisfied, the above stated three steps recur over generations.

1.4 Segmentation

Segmentation subdivides the patterns into disjoint regions or segments in such a way that the similar patterns may belong to same segments and the patterns in the different segments are dissimilar to each other. The patterns having the same characteristics are known as clusters/segments. Segmentation can be employed for localisation and detection of object specific features from both pictorial and nonnumeric data. The main motivator of these tasks lies in the effort to mimic human intelligence for extracting underlying objects perfectly. In our day-to-day life, uncertainty should be handled properly to represent the real-world knowledge. For example, image data evidences diverse uncertainty in content and structure. The main objective of segmentation is to detect the relevant and meaningful data by removing the redundant data embedded therein. The basis of any segmentation method relies on the variety of data representation, proximity measurement between data elements and grouping of data elements. For this, some metrics are usually applied to measure the similarity or dissimilarity between the patterns. Segmentation has been applied in different fields, such as defense, surveillance, robotic vision, machine learning, mechanical engineering, electrical engineering, artificial intelligence, pattern recognition, medical sciences, remote sensing, economics to name a few.

A pattern [87, 88], or an abstract representation of the dataset, is represented as $\mathbf{x}$ and can be denoted as $\mathbf{x} = (x_1, x_2, \ldots, x_d)$, where d signifies the number of features to represent the pattern. The pattern set is referred to as $\wp = \{\mathbf{x}_1, \mathbf{x}_2, \ldots, \mathbf{x}_n\}$ in which the ith pattern of this set is denoted as $\mathbf{x}_i = (x_{i,1}, x_{i,2}, \ldots, x_{i,d})$. The pattern set to be segmented can be presented using a pattern matrix, $\mathbf{X}_{n \times d}$. On the basis of some properties, a segmentation algorithm tries to derive a segment $\mathbf{S} = (S_1, S_2, \ldots, S_K)$ of K classes. A segmentation algorithm should follow the following properties [87, 88].

1. $S_i \neq \emptyset$ for $i = 1 \ldots, K$ i.e. each segment should contain at least one pattern.
2. $S_i \cap S_j = \emptyset$ for $i = 1 \ldots, K, j = 1 \ldots, K$ and $i \neq j$, i.e. the segments must be disjoint.
3. $\bigcup_{i=1}^{K} S_i = \wp$, i.e. each pattern should be included in a segment.

Different techniques have been applied to segment a dataset using all of the afore-mentioned properties. Proper fitness functions have been used to determine the validity of the partitioning. Mainly, segmentation algorithms can be segregated into three types, such as, supervised, semi-supervised and unsupervised segmentation algorithms. In the supervised segmentation algorithms, the number of desired partitions and labelled datasets are generally predefined. Moreover, in these types of segmentation algorithms, it has been attempted to keep the number of segments small and the datapoints are allotted to segments using an idea of closeness by resorting to a given distance function. These algorithms are employed to predict the correct partition of the disjoint subset of the dataset. In the semi-supervised segmentation algorithms [89] we have small amount of knowledge of the segmentation process that usually consists of either pairwise constraints between data items or class labels for some items. Basically, a limited form of guidance is used in these type of algorithms because the available knowledge about the target classification of the items is fully provided. The semi-supervised segmentation algorithms can be classified into two categories [89], such as, constraint-based methods and distance-based methods. In constraint-based methods [89], a modified distance function is furnished to incorporate the knowledge with respect to the classified examples and a traditional clustering algorithm is employed to cluster the data. Distance-based methods [89], on the other hand, modify the segmentation algorithm itself so that some previously supplied constraints or labels can be used to bias the search for an appropriate segmentation. Different ways have been employed for this, like, by including in the cost function a penalty for lack of compliance with the specified constraints [90], by performing a transitive closure of the constraints and using them to initialise clusters [91], or by requiring constraints to be satisfied during cluster assignment in the clustering process [92]. At the end, the unsupervised algorithms do not possess any a priori information about the labelled classes, decision-making criterion for optimisation or number of desired segments beyond the raw data or grouping principle(s) on the basis of their data content.

Image segmentation [87, 93–96], a major application areas of segmentation, is a critical and important component for analysis and understanding of image information, video and computer vision applications. This process is one of the major steps in image processing and analysis because it determines the quality of the processed image. A segmented image labels each object in such a way that it provides the description of the original image. The segmented image can be interpreted by the system easily to handle the image. Ultimately, image segmentation means the clustering of the pixels in an image into its constituent segments on the basis of some features such as intensity, shape, colour, position, texture and homogeneity. In a segmented image, the nonoverlapping regions are homogeneous and the union of any two adjacent regions is heterogeneous. Most of the segmentation algorithms are designed on the basis of the discontinuity and similarity of the intensity levels of an image. For a successful classification, some a priori knowledge or/and assumptions about the image have been usually expected in order to differentiate the features. The application areas of the image segmentation are in the fields of feature extraction, object recognition, machine vision, satellite image processing, medical imaging, biometric measurements, astronomical applications to name a few.

1.5 Role of Optimisation

Optimisation refers to a selection process of the most feasible solution from a set of acceptable solutions of a certain problem. This type of selection is done on the basis of some predefined single or multiple criteria. Basically, optimisation handles the problems in which one or more objectives of those problems are minimised or maximised and those objectives are functions of some real or integer variables. The proper values of real or integer variables within an allowed set are selected to solve the optimisation problems in a systematic way. Ultimately, the main objective of any type of optimisation is to derive the best value on the basis of some objective function within a definite domain. An optimisation function can be represented as follows [97]:

Suppose a function $f : A \rightarrow B$ from some set A to the set of real numbers B, the objective is to select an element $\bar{x}_m$ in A such that $f(\bar{x}_m) \geq f(\bar{x})$, $\forall \bar{x} \in A$ (maximisation) or such that $f(\bar{x}_m) \leq f(\bar{x})$, $\forall \bar{x} \in A$ (minimisation).

In this representation, the combination of operators like equalities or inequalities, and constraints is denoted by the Euclidean space B^n. A, the subset of B^n, must satisfy the entities of B^n. The cost function/objective function/energy function is denoted as the function f. A is the search space and the feasible or candidate solutions are the elements of A. An optimal solution is one of the feasiable solutions that is derived by optimising the objective function.Single-objective based optimisation problems are very much common in most of the research and application areas. In the real-world scenario, any optimisation problem considers as a single objective, although the most real-world problems are based on more than one objective.

Multi-objective optimisation [98–102] is considered as finding of one or more optimal solutions as the optimisation problem optimises more than one objective function with respect to a set of certain constraints. MOO is also known as multiattribute or multicriteria optimisation. In any MOO problem, the objective functions are typically dependent to each other. An optimisation problem falls short of being considered as a MOO if the optimisation of one objective guides the automatic optimisation of the other. The application areas of MOO is vast encompassing network analysis, bioinformatics, finance, automobile design, the oil and gas industry, product and process design, etc. A brief description of the single and multi-objective optimisation is narrated in the following subsections.

1.5.1 Single-Objective Optimisation

The main objective of single-objective (SO) optimisation is to detect the best solution on the basis of the minimisation or maximisation function. In this method, it may happen that different objectives are combined into a single objective to get a compact objective. This methods are verye much helpful for getting a clear viewe of the nature of the problem. To deal with different objectives, SO optimisation method is unable to supply a set of alternative solutions. In a broad sense, the single-objective optimisation techniques are segregated into three categories [97].

- Calculus-based techniques
- Enumerative techniques
- Guided random techniques

In calculus-based methods, alias, numerical methods, the optimisation problem is satisfied by the solutions on the basis of sufficient and necessary conditions. Each and every point of the finite, or discretised infinite, search space is evaluated by the enumerative techniques in order to arrive at the optimal solution. Enumerative methods in combination with the search space to guide the search to potential regions of the search space is known as guided random search techniques.

1.5.2 Multi-objective Optimisation

We have to take decisions to solve different problems in our daily life, consciously or subconsciously. These decisions may be simple or complex. When people go to the market and buy a product, it is very easy to choose a product if they consider only a single criteria for selection. Suppose, a person wants to buy a product with minimum expense. He can select the product quite easily as he is not considering the quality of the product. On the opposite side, he has to pay more if he considers quality of the product. He will get a single solution, i.e. either the product with minimum cost but low quality or the product having best quality with high cost. This scenario may become more complex if he considers both the criteria to buy that same product. In that case, the person may get more than one feasible solutions and he has to choose one among them. So, the degree of complexity may increase if more than two criteria are considered together to solve a problem. Ultimately, the desired result in different types of decision-making problems is achieved by optimising one or more criteria. In the same manner, the growth of optimisation problems in computer science becomes a great challenge. The problem can be intensified by the fact if several incommensurable and competing constraints of that problem have to be optimised simultaneously. To solve the real-life problems, several objectives must be optimised simultaneously in order. It is observed in many problems that the selected objectives are conflicting with each other. It is quite difficult to compare one solution with another one as there is no accepted definition of optimum in this case. Several objectives have to be optimised simultaneously in order to solve that problem. These types of problems are recognised as multi-objective optimisation problems (MOOPs). In general, multiple solutions will be obtained from these problems, each of which is considered acceptable and equivalent based on the relative importance of the unknown objectives.

Multi-objective optimisation [98–102] is considered as obtaining one or more optimal solutions as the optimisation problem requires more than one objective function. Mathematically, the multi-objective optimisation (MOO) can be defined as [98–105]

$$\text{Optimise the vector function } \overline{f}(\overline{x}) = [\overline{f_1}(\overline{x}), \overline{f_2}(\overline{x}), \dots, \overline{f_k}(\overline{x})]^T \qquad (1.12)$$

subject to the m inequality constraints [98–104]

$$g_i(\overline{x}) \geq 0, i = 1, 2, \dots, m \qquad (1.13)$$

and the p equality constraints [98–104]

$$h_i(\overline{x}) = 0, i = 1, 2, \dots, p, \qquad (1.14)$$

where k is the number of objective functions $f_i : R^n \rightarrow R$. The vector of n decision variables $\overline{x} = (x_1, x_2, \dots, x_n)^T$ is resultant solution, $\overline{x}$, where $x_i^L \leq \overline{x} \leq x_i^U, i = 1, 2, \dots, n$. Each decision variable x_i is limited within the corresponding lower bound (x_i^L) and upper bound (x_i^R).

The feasible region $\mathcal{F}$ which contains all the admissible solutions is induced by the constraints given in Eqs. 1.13 and 1.14. Any solutions outside this region will not be admitted as it violates one or more constraints. The optimal solution in $\mathcal{F}$ is denoted by the vector $\overline{x}$. However, it is difficult to establish the definition of the optimality in the context of multi-objective optimisation. Hardly a solution, represented by single vector $\overline{x}$, represents the optimum solution to all the objective requirements together [106].

The foremost problem considering multi-objective optimisation (MOO) is that there is no accepted definition of optimum in this case and therefore it is difficult to compare one solution with another one. Solutions to a multi-objective optimisation method are mathematically presented in terms of non-dominated solutions or *Pareto-optimal solutions*. The discussion of the Pareto optimality is the main concern in the domain of multi-objective optimisation. From the standpoint of minimisation

Fig. 1.1 Example of
dominance using five
solutions

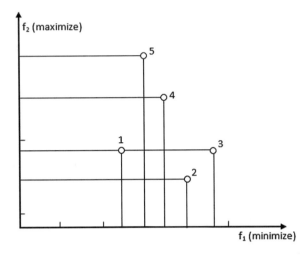

problem, the formal definition of domination may be defined as [98, 99, 103, 104]:
a decision vector $\overline{x_1}^*$ is declared as dominate a decision vector $\overline{x_2}^*$ (also denoted as
$\overline{x_1}^* \prec \overline{x_2}^*$) iff

$$\forall i \in 1, 2, \ldots, k, \; f_i(\overline{x_1}*) \geq f_i(\overline{x_2}^*) \tag{1.15}$$

and

$$\forall j \in 1, 2, \ldots, k, \; f_j(\overline{x_1}*) > f_j(\overline{x_2}^*) \tag{1.16}$$

Let us consider a two-objective optimisation problem with five different solutions
depicted in the objective space to explain the above mentioned concept and it is shown
in Fig. 1.1. Let us consider that one objective function f1 needs to be minimised while
another objective function f2 needs to be maximised. In this figure, five solutions with
different values of the objective functions are shown. It is very difficult to derive the
best solution from the above example with respect to both objectives as both objective
functions have same importance. However, the definition of domination is applied to
decide the better solutions among the given solutions in terms of both objectives. It is
obvious from the figure that solution 1 dominates solution 2 as the former performs
better than the latter with respect to both objectives. Again solution 3 is dominated
by the solution 4 but solution 4 is dominated by solution 5. Now, consider solution
1 and 5, they are non-dominated to each other as solution 1 is better than solution
5 in respect of f1 but the performance of solution 1 is not better than solution 5
interms of f2. So, we can conclude that solution 1 and 5 are non-dominated to
each other. Thus, different solutions with multiple objectives can be compared with
the help of domination concept. Ultimately, we may infer that there may be three
possibilities to check the dominance between two solutions 1 and 2, i.e. (i) solution
2 is dominated by solution 1 (ii) solution 2 dominated solution 1 (iii) solutions 1
and 2 are non-dominated to each other. It is to be mentioned that the dominance

relation also follows different binary relation properties, like, reflexive, symmetric, antisymmetric and transitive. These four relation properties are discussed as follows:

- **Reflexive**: The dominance relation is non-reflexive as a solution p does not dominate itself. The second condition of the definition is also not satisfied in this case [97, 98].
- **Symmetric**: Symmetric relation is not applicable for the dominance relation, because if p1 dominates p2, this does not mean that p2 dominates p1. Basically, the opposite is true and thus it may be concluded that the dominance relation is asymmetric [97, 98].
- **Antisymmetric**: The dominance relation cannot be antisymmetric as the dominance relation is not symmetric [97, 98].
- **Transitive**: The dominance relation is transitive. This is because, if p dominates q and q dominates r, then p dominates r. The dominance relation also has another interesting property that if a solution p does not dominate solution q, this does not imply that q dominates p [97, 98].

In addition, we can say $\overline{x_1}^*$ covers $\overline{x_1}^*$ if $\overline{x_1}^* \succ \overline{x_2}^*$ or $\overline{x_1}^* = \overline{x_2}^*$. All decision vectors which are not dominated by any other decision vector are called *non-dominated* or *Pareto optimal*. Without worsening at least one other objective, a Pareto-optimal solution cannot be improved with respect to anyone objective.

Hence, the Pareto-optimal set is mentioned as the set of all feasible non-dominated solutions in $\overline{x}$ and the *Pareto front* is presented as the corresponding objective function values in the objective space for a given Pareto-optimal set. Based on the dominance relationship, the non-dominated set is categorised into two types, i.e. *strongly non-dominated* and *weakly non-dominated* set [98]. The strongly non-dominated set is determined as a solution $\overline{x_1}^*$ which strongly dominates a solution $\overline{x_2}^*$, that means the solution $\overline{x_1}^*$ is strictly better than solution $\overline{x_1}^*$ with respect to all constraints [98, 99, 103, 107]. On the contrary, the solutions those are not strongly dominated by any other feasible solutions of the set $\Re$ is known as weakly non-dominated set [98, 103, 107].

The non-dominated set of solutions can be derived from a population of solutions using the previously stated concepts. The following procedure can be employed to deduce the set of non-dominated solutions from a given set P of N solutions. Each solution has M (>1) objective function values [98, 107].

1. Create a non-dominated set $P' = \emptyset$ and solution counter $i = 1$.
2. Check if solution j dominates solution i for a solution $j \in P$ (but $j \neq i$). If yes, go to Step 4.
3. $j = j+1$. If more solutions are left in P and go to Step 2; otherwise set $P' = P' \cup (i)$.
4. $i = i+1$. If $i \leq N$, go to Step 2; otherwise stop and declare P' as non-dominated set.

Population-based evolutionary algorithms such as genetic algorithms have tremendous advantage to solve multi-objective optimisation problems (MOOP). They are categorised as aggregating approaches, population-based non-Pareto approaches and Pareto-based non-elitist approaches [97, 98].

- Plain aggregating approaches: Plain aggregating approaches are also known as classical approaches as they do not work on the principle of evolutionary algorithms. Few non-dominated solutions are applied to find out the particular solution. During the optimisation process, some preferences may or may not be attached on the relative importance of each objective. Basically, these preferences are required to convert the MOOP into a single-objective optimisation problem, favouring a single Pareto-optimal solution. However, more than one Pareto-optimal solutions can be derived by changing the weights during optimisation. These types of optimisation algorithms are classified into four categories [98].

  No-preference methods   : In these methods, single optimal solution is obtained by a heuristic search, without having any information and giving any importance to the objectives. Multiple Pareto-optimal solutions are not derived for ensuring no preference to any objectives.

  Posteriori methods   : These methods apply preferences on each objectives to generate a set of Pareto-optimal solutions in iterative way.

  A priori methods   : These methods apply more information about preferences of objectives to derive one preferred Pareto-optimal solution.

  Interactive methods   : These methods employ the preference information increasingly during the optimisation process.

  Weighted sum approach, goal-programming based approach, goal attainment-based approach, ϵ-Constraint approach, etc. are some of the well-known plain aggregating approaches or classical approaches.

- Population-based non-Pareto approaches: In these types of approaches, evolutionary algorithm (EA) plays a vital role as EA starts its process with a population of solutions. These types of approaches are very much advantageous to find as many Pareto-optimal solutions as possible. The conventional EAs are made some algorithmic changes to achieve the Pareto-optimal solutions. Different parameters of the classical approaches, like, weight vectors, ϵ vectors, target vectors, etc. are not needed in these types of approaches. On the contrary, the EA's population-oriented approach gives equal importance to all non-dominated solutions in a population and tries to preserve a diverse set of multiple non-dominated solutions simultaneously [98]. A child population is generated by preserving multiple good solutions. Ultimately, the process will generate the resultant Pareto-optimal front and with a good spread. The ancient multi-objective evolutionary algorithm, vector evaluated GA (VEGA), was proposed by David Schaffer in 1984 [108, 109]. Other than VEGA, lexicographic ordering, game theory-based approach, contact theorem based approach, nongenerational GA based approach, etc. are some of the population-based non-pareto approaches [97].

- Pareto-based non-elitist approaches: In these types of approaches, dominated solutions in the current population are isolated from the non-dominated solutions in that population. A non-dominated ranking procedure is introduced by Goldberg [73, 98] to determine the fitness of the individuals. It works on the concept of Pareto dominance and each non-dominated solutions are given equal reproduction probabilities. The ranking procedure is done iteratively. In this process, rank 1 is assigned to all non-dominated solutions and they are eliminated from the population. After that, the next non-dominated solutions in the remaining population are assigned rank 2 and so on. Based on this concept, the multi objective GA (MOGA) [110], Niched Pareto GA (NPGA) [111], Non-dominated sorting GA (NSGA) [112] are few examples of Pareto-based non-elitist approaches.

- Pareto-based elitist approaches: A new concept of *elitism* introduced in these types of approaches in comparison to the previously stated approaches. An elite-preserving operator is applied to preserve the elites of a population so that they can get direct entry in the next generation. The concept of elitism can be stated in simple manner. After crossover and mutation operators, the offsprings are compared with both of their parents. After that, the best two solutions are selected among the four parent-offspring solutions. So in broad sense, both parent and offspring population are combined after the offspring population generation. That means, the best N members are selected from the $2N$ solutions. This population will be the next population of the next generation. Based on this concept, Strength Pareto evolutionary algorithm (SPEA) [113], Strength Pareto evolutionary algorithm 2 (SPEA2) [114], Pareto archived evolutionary strategy (PAES) [115], Pareto envelope-based selection algorithm (PESA) [115], Pareto envelope-based selection algorithm-II (PESA-II) [116], Elitist non-dominated sorting GA (NSGA-II) [102], etc. are developed.

1.6 Organisation of the Book

In the present scope, we have tried to provide a glimpse of different approaches including Soft and Hybrid approaches used in image segmentation. We shall discuss some more important issues subsequently in details in the chapters ahead.

Apart from the present Chapter, Chap. 2 provides the direction of research in the field of image segmentation with a bit more details of several classical and non-classical approaches for image segmentation.

Chapter 3 offers an overview of the mathematical foundations needed for the technical discussion in subsequent chapters of the book. It describes a scenario of the multilevel greyscale image segmentation using conventional multilevel sigmoidal (MUSIG) activation function. The choice of such a function is very significant. It uses equal and fixed class responses, thereby ignoring the heterogeneity of image information content. In this chapter itself, we shall also discuss a novel approach for generating optimised class responses of the MUSIG activation function, named optimised MUSIG (OptiMUSIG) activation function. This is done to ensure that the image content heterogeneity can be incorporated in the segmentation procedure. Different types of objective functions are used to measure the quality of the segmented images in the proposed GA based optimisation method.

Colour image segmentation happens to be the logical but challenging extension of multilevel greyscale image segmentation, which comes under the purview of Chap. 4. We present a parallel version of the OptiMUSIG activation function in this chapter and focus on the segmentation of true colour test images, based on all possible combination of colour intensity features. A parallel version of the OptiMUSIG (ParaOptiMUSIG) activation function is proposed in this chapter with the optimised class responses for the individual features with a parallel self-organising neural network (PSONN) architecture to segment true colour images. Different standard objective functions are applied to measure the quality of the segmented images in the proposed GA based optimisation method.

In the subsequent Chapter, viz. in Chap. 5, we strive to evolve refined and pruned versions of single-objective based OptiMUSIG activation function for multilevel greyscale image segmentation. This Chapter proposes a self-supervised image segmentation method by a Multi-Objective GA (MOGA) based optimised MUSIG (OptiMUSIG) activation function applied on Multi-Layered Self-Organising Neural Network (MLSONN) architecture to segment multilevel greyscale intensity images. A Non-dominated Sorting Genetic Algorithm-II (NSGA-II) based optimised MUSIG (OptiMUSIG) activation function with the MLSONN architecture is also presented in this chapter to segment multilevel greyscale images. The quality of the segmented images is evaluated using different standard objective functions and some of these functions are employed to form the multiple objective criteria of the proposed multilevel greyscale image segmentation procedure.

We present a MOGA based parallel version of optimised MUSIG (ParaOptiMUSIG) activation function with the PSONN architecture to segment the colour images in Chap. 6. A NSGA-II based ParaOptiMUSIG activation function coupled with the Parallel Self-Organising Neural Network (PSONN) architecture is also available in this Chapter for colour image segmentation. These activation functions are employed to overcome the shortcomings of the single-objective based ParaOptiMUSIG activation function. The quality of the segmented images is evaluated using different standard objective functions and some of these functions are employed to form the multiple objective criteria of the proposed colour image segmentation procedure.

A new GA based automatic image clustering technique is discussed in Chap. 7. A fuzzy intercluster hostility index is proposed here to delineate the different cluster centroids. A comparative study of the performances of the proposed algorithm and the Automatic Clustering with Differential Evolutionary (ACDE) is also presented.

Chapter 2
Image Segmentation: A Review

2.1 Introduction

This chapter intends to provide a brief review of different image segmentation techniques. Image segmentation techniques are broadly classified into two categories, viz. classical and non-classical approaches. Most of the classical approaches depend on filtering and statistical techniques. Methods in this direction engage thresholding techniques, edge detection or boundary based techniques, region-based techniques, morphological techniques, normalised cut—a graph theoretic approach, k-means approaches, etc. The non-classical approaches consisting of the fuzzy-neuro-genetic paradigm or its variants are contributed with features for real time applications.

This chapter presents a brief survey of the aforestated trends in image segmentation techniques.

2.2 Classical Approaches to Image Segmentation

Several classical approaches of image segmentation and analysis have been reported in the literature [93, 94]. One of the important techniques for image segmentation is the thresholding technique. This technique deals with the pixel value of an image, such as grey level, colour level, texture, to name a few. In this technique, it is presumed that the adjacent pixel values lying within a certain range belong to the same class [93]. The main objective of this technique is to determine the threshold value of an image. The pixels whose intensity values exceed the threshold value are assigned in one segment and the remaining to the other. The image segmentation approaches using histogram thresholding techniques can be found in the literature [17, 117]. The histogram thresholding technique based on the similarity between grey levels has been demonstrated in [118]. In this article, a mathematical model has been derived using the fuzzy framework. Image segmentation using entropy-based thresholding

© Springer International Publishing AG 2016
S. De et al., *Hybrid Soft Computing for Multilevel Image and Data Segmentation*, Computational Intelligence Methods and Applications,
DOI 10.1007/978-3-319-47524-0_2

techniques have been demonstrated in [117, 119]. Afrin and Asano [120] proposed a grey level thresholding algorithm on the basis of the relationship between the image thresholding problem and cluster analysis. The cluster similarity measure, applied to determine the threshold value, is the main criteria of this algorithm. The criterion-based concept to select the most suitable greyscale as the threshold value has been employed for many thresholding techniques. The Otsu's [121] thresholding method, one of the oldest methods, has been utilised to analyse the maximum separability of classes. This method has been used to determine the goodness of the threshold value that includes evaluating the heterogeneity between both classes and the homogeneity within every class. In [122], a criterion function has been applied in the threshold selection method based on cluster analysis. The criterion function is generated on the basis of the histogram of the image as well as the information on spatial distribution of pixels. This function has been applied to determine the intra-class similarity to attain the homogeneous and heterogeneous class differentiated by the inter-class similarity. A colour image segmentation method by pixel classification in a hybrid colour space which is adjusted to analyse the image, has been presented by Vandenbroucke et al. in [123]. Tan and Isa [124] proposed a hybrid colour image segmentation technique based on histogram thresholding and fuzzy c-means algorithm. In this technique, the histogram thresholding method has been applied to determine the consistent regions in the colour image. This step is followed by fuzzy c-means algorithm for the betterment of the compactness of the segments. The drawback of the histogram thresholding technique is that this technique overlooks all the spatial relationship information of the images except the pixel intensity. Hence, this technique cannot be efficiently applied for blurred images or for multiple image component segmentation.

Another classical image segmentation technique is the edge detection or boundary based technique. In these techniques, it is assumed that the pixel values change abruptly at the edges between two regions in the image [93]. Few simple edge detector operators such as the Sobel [93] or Roberts [93] operators or more complex ones such as the Canny operator [93] have been applied in these techniques. The derived discontinuous or over-detected edges are obtained by most of the existing edge detectors. However, it is desirable that the output of the actual region boundaries should be closed curves. The closed region boundaries can be prevailed by some post-procedures, such as edge tracking, gap filling, smoothing and thinning. Different edge detection techniques for image segmentation have been reported in different literatures [125, 126]. Sumengen and Manjunath [125] presented a multiscale edge detection technique for segmenting natural images. This approach is quite efficient to annihilate the necessity of explicit scale selection and edge tracking methods. An edge detection and image segmentation technique by designing a vector field has been proposed in [126]. In this method, the location where the vectors diverge from each other in opposite directions has been noted as the edges. A novel automatic image segmentation technique is presented by integrating an improved isotropic edge detector and a fast entropic thresholding technique in [127]. The main geometric structures of an image are evaluated by the derived colour edges. The initial seeds for seeded region growing are determined from the centroids between the adjacent edge regions. The centroids of the generated homogeneous image regions are then used to substitute these seeds by integrating the required additional pixels step by step. The main drawbacks of the edge detection technique are that these do not work well when images have many edges because in that case the segmentation method produces an over segmented result, and it cannot easily identify a closed curve or boundary. However, the global and continuous edges have been detected by the edge-based methods quite efficiently. An automated model-based image segmentation is presented by Brejl and Sonka [19, 128]. In this literature, a training set of two types of models, objects shape model and border appearance model, are maintained for segmentation. The image segmentation is done in two steps. The approximate location of the object of interest is determined in the first step, and in the next step that object is segmented using the shape-variant Hough transform method.

Region-based techniques have been widely used to solve image segmentation problems. The working principle of this technique is that it is assumed that the adjacent pixels in the same region possess similar visual features such as grey level value, colour value or texture. Split and merge [93], a well-known technique for region-based image segmentation, is very much dependent on the selected homogeneity criterion [129]. In region-growing or merging techniques, the input image is first fitted into a set of homogeneous primitive regions. After that, similar neighbouring regions are merged according to a certain decision rule using an iterative merging process [93]. On the other hand in splitting techniques, the entire image is initially regarded as one rectangular region. This process is terminated when all the regions of the image are homogeneous as each heterogeneous image region of the image is divided into four rectangular segments in each step. In split-and-merge techniques, after the splitting stage a merging process is applied for unifying the resulting similar neighbouring regions [93]. Basically, region-based techniques are greedy techniques following the intuitive process to continuously merge/split the regions based on some criteria. A heuristic segmentation approach based on region-merging method has been applied on oversegmented images in [130]. In this method, the characteristics of each pixel of the input image have been used to guide the merging process. Adams and Bischof [131] proposed a hybrid method of seeded region growing (SRG) for image segmentation. In this method, the SRG technique is controlled by a number of initial seeds without tuning the homogeneity parameters. The main objective of this method is that each connected component of a region is assigned with one of the seeds to find an accurate segmentation of images into regions. The regions are constructed by merging a pixel into its nearest neighbouring seed region. An automatic seeded region-growing algorithm for colour image segmentation has been presented in [132]. Gomez et al. [133] proposed an automatic seeded region-growing algorithm, named ASRG-IB1, to segment colour and multispectral images. In this method, the histogram analysis for each band is applied to generate the seeds automatically and an instance-based learning process is used as distance criterion. However, the problem of seeded region-growing technique is the automatic selection of the initial seeds to provide an accurate image segmentation. Lie et al. [134, 135] presented a variant of the level set method to segment an image. In this method, a piecewise constant level set formulation is introduced for identifying the separating regions in different phases. Storage capacity is also benefited by this approach. An unsupervised stochastic model based method to segment an image is presented by Guo and Ma [136, 137]. In this method, parameter estimation and image segmentation is done on the basis of Bayesian learning. A competitive power-value based approach is introduced to segment the images into different classes.

Mathematical morphology [93, 138] dealing with geometrical features, is a well-suited technique for segmentation purposes. Basically from the scientific perspective, the word morphology indicates a particular area of biology that handles the form and structure of animals and plants. In image processing, mathematical morphology has been applied for analysing and representing the geometric structures inherent within an image, such as boundaries, skeletons and convex hulls. Two basic operations of the mathematical morphology are erosion and dilation. These operators create an opening operator in which the erosion operator is applied to an image followed by the dilation operator. Watershed transformation is a powerful morphological tool for image segmentation [93]. Gauch [139] applied watershed transformation on gradient images to obtain an initial segmentation. The scale-based watershed region hierarchies are generated through multiscale analysis of intensity minima in the gradient magnitude. This approach usually yields the oversegmented result of an image. Belaid and Mourou [140] presented an approach to overcome this oversegmentation by the watershed transformation in combination with a topological gradient approach. The multiscale morphological approach [141] has been applied quite efficiently to segment grey level images consisting of bright and dark features of various scales. The potential regions at various scales have been identified at the first step followed by segment validation which has been done using three criteria, such as growing, merging and saturation. An unsupervised morphological clustering method [142] for colour image segmentation has been applied to analyse bivariate histogram. In this method, a segmentation fidelity and segmentation complexity based energy function is presented to measure the multiscale image segmentation quality. The merging of the oversegmented regions is also performed based on that energy function.

Out of the existing image enhancement procedures, the normalised cut—a graph-theoretic approach has become very popular over the years for addressing the problem of image segmentation. Shi and Malik [143] proposed a general image segmentation approach based on normalised cut by solving an eigensystem of equations. In this method, the normalised cut has been applied to partition the image to handle the image segmentation problem. The total likelihood within the regions has been measured by the normalised cut criterion. Images consisting of texture and non-texture regions can be segmented based on local spectral histograms and graph partitioning methods [144–146]. Yang et al. [145] proposed an image segmentation method referred to as the First Watershed Then Normalised Cut (FWTN) method, based on watershed and graph theory. The mean shift analysis (MS) algorithm has been proposed by Comaniciu and Meer [146] to determine the exact estimation of the colour cluster centres in colour space. The MS clustering method [147] can also be employed for designing an unsupervised multiscale colour image segmentation algorithm. In this algorithm, a minimum description length criterion has been applied to merge the resultant oversegmented images. Mean shift (MS) and normalised cut can be applied to segment colour images [148] as well. The discontinuity property of images can be maintained by the mean shift algorithm to form segmented regions. Finally, the normalised cut can be used on the segmented regions to reduce the complexity of the process. This segmentation algorithm provides superior outcome as it is applied on the adjacent regions instead of image pixels. Wang and Siskind [149] developed a cost function, named as ratio cut, based on the graph reduction method to segment colour images efficiently. This cost function works efficiently with region-based segmentation approaches as well as pixel-based segmentation approaches. A graph partitioning approach for segmenting colour images has been demonstrated in [150]. In this method, the colour images are oversegmented using the watershed method and the final segmented images are derived by a graph structure based merging method. A hierarchical structure, named the binary partition tree [151], has been applied for filtering, retrieval, segmentation and image coding purposes. Malik et al. [136, 152] introduced a graph partitioning method to segment an image. In this method, an algorithm is presented for partitioning greyscale images into different regions on the basis of brightness and texture of the image. A graph theoretic framework of normalised cuts is used to divide the image into the regions of coherent texture and brightness. Belongingness of two nearby pixels to the same region is measured by this graph theoretic framework.

The k-means algorithm is one of the simplest and most popular iterative clustering algorithm [87, 153]. The procedure follows a simple and easy way to classify a given dataset to a certain number of clusters. The cluster centroids are initialized randomly or derived from some a priori information. Each data in the dataset is then assigned to the closest cluster. Finally, the centroids are reassigned according to the associated data point. This algorithm optimises the distance criterion either by maximising the intercluster separation or by minimising the intra-cluster separation. The k-means algorithm is a data-dependent greedy algorithm that may converge to a suboptimal solution. An unsupervised clustering algorithm, using the k-means algorithm, is presented by Rosenberger and Chehdi in [154]. An improved k-means algorithm for clustering is presented in [155]. In this method, each data point is stored in a kd-tree and it is shown that the algorithm runs faster as the separations between the clusters increase.

Most of the classical approaches discussed above, need some *a priori* knowledge regarding the image data to be processed either in the form of the underlying intensity distribution or about appropriate parameters to be operated upon. The other approaches based on the soft computing methods, however, operate on the underlying data regardless of the distribution and operating parameters.

2.3 Soft Computing Approaches to Image Segmentation

Soft computing techniques, viz. neural network, fuzzy logic and genetic algorithms, have been applied for image segmentation extensively in many literatures. A detailed survey of image segmentation using soft computing techniques is depicted in the following subsections.

2.3.1   Neural Network Based Image Segmentation

Segmentation and clustering of image data followed by the extraction of specified regions can be accomplished by neural networks due to the inherent advantages of adaptation and graceful degradation offered by them [39, 50, 156–160]. Kohonen's self-organising feature map (SOFM) [39] is a competitive neural network used for data clustering. Jiang and Zhou [157] presented an image segmentation method using the SOFM. The pixels of an image are clustered with several SOFM neural networks and the final segmentation is obtained by grouping the clusters thus obtained. The Hopfield's network [57, 161], is a fully connected network architecture with capabilities of auto-association, proposed in 1982. A photonic implementation of the Hopfield network using an optical matrix-vector multiplier is applied to binary object segmentation [162]. An image is segmented by the parallel and unsupervised approach using the competitive Hopfield neural network (CHNN) [128, 163]. In this work, the pixel clustering is done on the basis of the global information of the grey level distribution and a winner-take-all algorithm is employed to emulate the state transitions of the network. Kosko [60, 164] introduced the bi-directional associative memory (BAM) model which evidences similar network dynamics as the Hopfield model. Several image segmentation and pattern recognition processes by self-organising neural network architectures are reported in different literatures [165–167]. Image pixels are one of the criteria in the image segmentation arena for their global feature distribution [168, 169]. Chi [170] presented a new SOM-based k-means method (SOM-K) followed by a saliency map-enhanced SOM-K method (SOMKS) for natural image segmentation. In the preliminary step, the network is trained with intensity value of the pixels and different features of the colour space and after that the k-means clustering algorithm is applied to segment the image. The self-organising neural network (SONN) is introduced to segment textural image in the literature [136, 171]. The SONN architecture for textural image segmentation is based on the orientation of the textures within the images. The binary objects from a noisy binary colour image can be extracted quite efficiently using a single multilayer self-organising neural network (MLSONN) [3] by means of self-supervision. In this network, the standard backpropagation algorithm has been used to adjust the network weights with a view to arriving at a convergent stable solution. A layered version of the same has been proposed in [134, 172, 173] to deal with multilevel objects at the expense of a greater network complexity. However, the multilevel objects cannot be extracted from an image by this network as it uses the bilevel sigmoidal activation function. Bhattacharyya et al. [45, 49, 72, 174] addressed this problem by introducing a functional modification of the MLSONN architecture. They introduced a multilevel sigmoidal (MUSIG) activation function for mapping multilevel input information into multiple scales of grey. The MUSIG activation function is a multilevel version of the standard sigmoidal function which induces multiscaling capability in a single MLSONN architecture. The number of grey scale objects and the representative grey scale intensity levels determine the different transition levels of the MUSIG activation function. The input image can thus be segmented into different levels of grey using the MLSONN architecture guided by the MUSIG activation function. However, the approach assumes that the information contained in the images are homogeneous in nature, which on the contrary, generally exhibit a varied amount of heterogeneity. Neural networks have often been applied for clustering of similar data [136, 175] by means of a selection of underlying prototypes. Thereafter, each prototype is assigned to a pattern based on its distance form the prototype and the distribution of data. In this approach, the number of clusters is determined automatically. Pixel-based segmentation has been implemented in the literature [136, 176]. In this method, each pixel is assigned to a scaled family of differential geometrical invariant features. Input pixels are clustered into different regions using SOFM. SOFM has also been used for the unsupervised classification of MRI images of brain [134, 177, 178]. The MRI images are decomposed into small size by two-dimensional Discrete Wavelet Transform (DWT). The approximation image is applied to train the SOFM and the pixels of the original image are grouped into different regions.

Jiang et al. [179] presented a literature review on computer-aided diagnosis, medical image segmentation, edge detection, etc. by different neural network architectures. In this article, different techniques as well as different examples of neural network architectures are illustrated for medical image segmentation. A new version of the self-organising map (SOM) network, named moving average SOM (MA-SOM), is applied to segment the medical images [180]. SOFM has also been used to segment medical images through identification of regions of interest (ROI) [159, 160]. A pixel-based two-stage approach for segmentation of multispectral MRI images by SOFM is presented by Reddick et al. in [160, 172]. An unsupervised clustering technique for magnetic resonance images (MRI) is presented by Alirezaie et al. [128, 181]. The Self- Organising Feature Map (SOFM) artificial neural network is proposed to map the features in this method.

Neural networks are well known for their parallelism and ability of approximation, adaptation and graceful degradation [37–48]. Researchers applied neural networking techniques using these features for addressing the problem of colour image processing. Several works have been reported where different neural network architectures have been used in this regard [50, 182]. The competitive learning (CL) is applied for online colour clustering on the basis of the least sum of squares criterion [183]. This CL method efficiently converges to a local optimum solution for colour clustering. Wang et al. [184] proposed a two-stage clustering approach using competitive learning (CL) for fast clustering. The local density centres of the clustering data is identified by the CL and after that an iterative Gravitation Neural Network is applied to agglomerate the resulting codewords in a parallel fashion. The Kohonen self-organising feature map (SOFM) [157, 185, 186] also has the ability of retrieving the dominant colour content of images. The SOFM combined with the adaptive resonance theory (ART) has been applied for segmenting colour images in [185]. Self-organising maps (SOM) [157, 187] which are efficient to retrieve the dominant colour content of images have been widely employed to colour image segmentation arena [188]. An ensemble of multiple SOM networks [157] are employed for colour image segmentation in which segmentation is accomplished by the different SOM networks on the basis of the colour and spatial features of the image pixels. Finally, the desired output is derived by combining the clustered outputs. The primitive clustering results are generated by training the SOM on the basis of a training set of five-dimensional vectors (R, G, B, and x, y) in [186]. The isolated pixels are eliminated during the segmentation of the image which is done by merging the scattered blocks. SOFM is also applied to capture the dominant colours of images in [188]. These dominant colours are further merged to control the ultimate number of colour clusters. A neural network based optimal colour image segmentation method, which incorporates colour reduction followed by colour clustering, is proposed by Dong and Xie [189]. A SOFM network is applied for colour reduction to project the image colours which are represented in a modified $L* u * v$ space into a reduced set of prototypes. Finally, simulated annealing is used to find out the optimal clusters with a minimum computation cost in the SOFM-derived prototypes. In [190], a fast convergent network named Local Adaptive Receptive Field Self-organising Map (LARFSOM) is employed to segment colour images efficiently. Neural networks are also able to categorise content based images efficiently [191]. In this method, the region segmentation technique is employed to extract the objects from the background and after that the back propagation algorithm is applied in the neural network for feature extraction. SOFM-based neural network is also applied to segment multispectral Magnetic Resonance Images (MRI) images [192]. A hierarchically divisive colour map for colour image quantization with minimal quantization error has been proposed by Sirisathitkul et al. in [193, 194]. They used the Euclidean distances between the adjacent colours with principal axes along the highest variance of colour distribution, to divide a colour cell into two subcells. SOFMs are used to carry out clustering based on colour and spatial features of pixels. The clustered outputs are finally merged to derive the desired segmentation. A parallel version of the MLSONN (PSONN) architecture [195, 196] consisting of three independent and parallel MLSONNs (for component level processing) has been efficiently applied to extract pure colour images from a noisy background. Each network is used to process the colour image data at the R, G, B component levels. The real-world input colour images are provided as inputs through the source layer of the PSONN architecture [195, 196] and the sink layer finally fuses the processed colour component information into processed colour images. This architecture uses the generalised bilevel sigmoidal activation function with fixed and uniform thresholding. Bhattacharyya et al. [195] introduced a self-supervised parallel self-organising neural network (PSONN) architecture guided by the multilevel sigmoidal (MUSIG) activation function for true colour image segmentation. Since the utilised activation functions use fixed and uniform thresholding parameters, they assume homogeneous image information content.

2.3.2 Fuzzy Based Image Segmentation

Real life images exhibit a substantial amount of uncertainty. The varied amount of vagueness manifested in the colour image intensity gamut can be efficiently handled by fuzzy set theory and fuzzy logic. Fuzzy set theory and fuzzy logic [64–68] are very much familiar techniques in the field of image processing and image classification. A good literature survey of the colour image segmentation using fuzzy logic is presented in the literature [197, 198]. The fuzzy c-means (FCM) clustering algorithm [199] is the most popular method based on membership values of different classes. This technique is efficient in clustering multidimensional feature spaces. Each data point belongs to a cluster to a degree specified by a membership grade. The fuzzy c-means algorithm partitions a collection of n pixels $X_i, i = \{1, \ldots, n\}$ into c fuzzy groups, and finds a cluster centre in each group such that a cost function of dissimilarity measure is minimised. However, the FCM algorithm does not fully utilise the spatial information and it only works well on noise-free images. Ahmed et al. [172, 200] proposed a modified version of the objective function for the standard FCM algorithm to allow the labels in the immediate neighbourhood of a pixel to influence its labelling. The modified FCM algorithm improved the results of conventional FCM method on noisy images. A geometrically guided FCM (GG-FCM) algorithm is introduced in [201], based on a semi-supervised FCM technique for multivariate image segmentation. In this method, the local neighbourhood of each pixel is applied to determine a geometrical condition information of each pixel before clustering. Incorporating the spatial constraints to segmentation algorithms [200, 202] and imposing different features or dissimilarity index that is insensitive to intensity variations in the objective function of FCM [203, 204] are two different classifications of the modified FCM. Rezaee et al. [128, 205] introduced an unsupervised image segmentation technique. The segmentation process is the combination of pyramidal image segmentation with the fuzzy c-means clustering algorithm. The root labelling technique is applied to separate each layer of the pyramid into a number of regions. The fuzzy c-means is applied to merge the regions of the layer with the highest image resolution. The minimum number of objects is grouped by the cluster validity function. An algorithm has been proposed in [134, 206] to segment an image based on fuzzy connectedness using dynamic weights (DyW). Fuzzy connectedness is measured on the basis of the linear combination of an object-feature and homogeneity component using fixed weights.

Fuzzy set theory and fuzzy logic can incorporate the colour and spatial uncertainty and guide the segmentation process [207]. Fuzzy rules have been applied for the region dissimilarity function and the total merging process in [208, 209]. Other fuzzy homogeneity based colour image segmentation methods have been presented in the literature [193, 210]. Yang et al. [211] proposed an eigen-based fuzzy C-means (FCM) clustering algorithm to segment colour images. In this method, the eigenspace is formed by dividing the selected colour pixels of the image. colour image segmentation is achieved by combining the eigenspace transform and the FCM method. A fuzzy min-max neural network based colour image segmentation technique (FMMIS) is proposed by Estevez et al. [212] to detect the image artefacts. The minimum bounded rectangle (MBR) for each object, present in an image, is determined in this method and then the method grows boxes around starting seed pixels to delineate different object regions in the image. Fuzzy logic in combination with the seeded region-growing method is applied for colour image segmentation in [213]. The initial seed in this method is selected with the proposed fuzzy edge detection method which has been used to detect the connected edges. A colour image segmentation algorithm named, eigenspace FCM (SEFCM) algorithm, is efficient to segment the images that have the same colour as the pre-selected pixels [214]. In this method, the colour space of the selected colour pixels of the image is segregated into principal and residual eigenspaces.

Fuzzy techniques can be combined with neural network to segment the multidimensional images, viz. colour, satellite, multi-sensory images [215]. In this approach, a statistics of the features of the framework and the parameters has been generated by analysing different segmented images and this statistics is employed in the segmentation of different images taken under difficult illumination conditions. An incremental Weighted Neural Network (WNN), proposed by Muhammed [216], is employed for unsupervised fuzzy image segmentation. The topology of the input data set is preserved by the interconnected nodes in the proposed network. The fuzziness factor is proportional to the connectedness of the net. Finally, the segmentation procedure is carried out by the watershed clustering method and this process is also applied to determine the number of clusters.

2.3.3 Genetic Algorithm Based Image Segmentation

Genetic algorithms (GAs) [74, 217, 218], randomised search and optimization techniques guided by the principles of evolution and natural genetics, work on the collection of probable solutions in parallel rather than on the domain dependent knowledge. GAs are ideal for those problems that do not have any knowledge about the domain theories of the problem or difficult to formulate the problem. Near optimal solutions with an objective or fitness function are provided by the GAs. Due to generality of the GAs, they are applied to solve the image segmentation problem and only require a segmentation quality measurement criterion. Biologically inspired on principles of natural genetic mechanisms, such as population generation, natural selection, crossover and mutation are applied over a number of generations for generating potentially better solutions. An extensive and detailed work of image segmentation using GAs is depicted by Bhanu et al. [219]. Alander [220] presented and compiled a complete survey on GAs used in image processing. Several notable applications of GA based image segmentation approaches can be found in the literature [172, 219, 221, 222]. Yoshimura and Oe [220, 223] proposed an approach for the clustering of textured images in combination with GA and Kohonen's self-organising map. GA has been applied for the clustering of small regions in a feature space and used for developing the automatic texture segmentation method. A combined approach of genetic algorithm with the K-means clustering algorithm has been employed for image segmentation in [224]. In this method, a fitness function is proposed on the basis of texture features similarity [220]. The hierarchical GA has the ability to cluster an image with a predefined number of regions [134, 225]. An image segmentation algorithm is proposed using the fuzzy rule based genetic algorithm in [134, 226]. A

K-means clustering technique is used in this approach to reduce the search space before the application of the GA. Thereafter, GA is engaged to split the regions. Since, GAs have the capability to generate class boundaries in an N-dimensional data space, a set of nonredundant hyperplanes have been generated in the feature space to produce minimum misclassification of images in [227]. In [228], a three-level thresholding method for image segmentation is presented, based on the optimal combination of probability partition, fuzzy partition and entropy theory. A novel approach of fuzzy entropy is applied for the automatic selection of the fuzzy region of membership function [228, 229]. After that, the image has been translated into fuzzy domain with maximum fuzzy entropy and genetic algorithm has been implemented to determine the optimal combination of the fuzzy parameters. In this method, the thresholding is decided by executing the fuzzy partition on a two-dimensional (2-D) histogram based on fuzzy relation and maximum fuzzy entropy principle. An entropy function, named monotonic, is proposed by Zhao et al. [228, 230] and is evaluated by the fuzzy c-partition (FP) and the probability function (PF) to measure the compatibility between the PP and the FP. This function is also applied to determine the memberships of the bright, the dark and the medium approximately. The GAs in combination with the classical fuzzy c-means algorithm (FCM) is applied for colour image segmentation and the objective function of the FCM is modified by considering the spatial information of image data and the intensity inhomogeneities [231]. This image segmentation method does not require the prefiltering step to eliminate the background.

A novel approach of image segmentation by GA is presented in the literature [128, 136, 232]. Mean square error (MSE) is applied as the criteria to segment an image into regions. Shape and location of the regions are also used as the criteria in this method. An image segmentation method by pixon-based adaptive scale is presented in the literature [158, 233]. The pixon-based image model is a combination of a Markov random field (MRF) model under a Bayesian framework. The pixons are generated by the anisotropic diffusion equation. Feitosa et al. [193, 234] proposed a fitness function based on the similarity of resulting segments to target segmentation. The process is applied to minimise the parameter values by modifying the parameters of the region-growing segmentation algorithm. Manual computation is still required though computation is straightforward and intuitive. This method can easily be adapted to modify parameters of other segmentation methods. Image segmentation using fuzzy reasoning in combination with GAs is presented by Zhu and Basir [235]. In this approach, fuzzy reasoning is done in two steps, a denoising phase and a region-merging phase. The segmentation results are optimised by the genetic algorithms. Yu et al. [136, 236] introduced a image segmentation method using GA combined with morphological operations. In this method, morphological operations are used to generate the new generations.

Sun et al. [237] proposed a clustering technique using neural network and GA. In this method, genetic algorithm based maximum likelihood clustering neural networks (GAMLCNN) is applied for segmentation. A combination of genetic algorithm and wavelet transform based multilevel thresholding method is presented by Hammouche et al. [238] for image segmentation. The appropriate number of thresholds as well as the tolerable threshold value is determined by this method. The wavelet transform is applied to reduce the length of the original histogram and the genetic algorithm is employed to decide the number of thresholds and the threshold values on the basis of the lower resolution version of the histogram.

Medical image segmentation can be handled by the genetic algorithm though the medical images have poor image contrast and artefacts that result in missing or diffuse organ/tissue boundaries [239]. In this article, a detailed review of medical image segmentation using genetic algorithms and different techniques is discussed for segmenting the medical images. Zhao and Xie [240] also presented a good literature survey on interactive segmentation techniques for medical images. Lai and Chang [241] proposed a hierarchical evolutionary algorithms (HEA), mainly a variation of conventional GAs, for medical image segmentation. In this process, a hierarchical chromosome structure is applied to find out the exact number of classes and to segment the image into those appropriate classes. Automatic clustering using differential evolution (ACDE) is applied to extract the shape of the tissues on different type of medical images automatically [242, 243]. Ghosh and Mitchell [242, 244] applied genetic algorithm for segmenting the two-dimensional slices of pelvic computed tomography (CT) images automatically. In this approach, the segmenting curve is used as the fitness function in GA. A hybridized genetic algorithm (GA) with seed region-growing procedure is employed to segment the MRI images [242, 245]. A combined approach of neural network and genetic algorithm is presented in [134, 246] to segment an MRI image.

A genetic algorithm based colour image segmentation method is proposed by Farmer and Shugars [247]. GAs have been used in image segmentation for the optimization of relevant parameters in the existing segmentation algorithms [247]. Image segmentation can be easily devised as an optimization problem to reduce the complexity and redundancy of image data. Zingaretti et al. [193, 248] applied GAs to solve the unsupervised colour image segmentation problem. The most important feature of this method is that it performs multi-pass thresholding. In this process, the genetic algorithm adopts different thresholds during each pass to segment a wide variety of non-textured images successfully. Gong and Yang [249] used quadtrees to represent the original and the segmented images. A two- pass system has been defined in this method, where, in the first pass, a Markov Random field based energy function has been minimised and in the second pass, genetic algorithm has been used for final segmentation. In this process, the multi-resolution framework for using the different strategies at different resolution levels and to accelerate the computation has been implemented using the quadtree. This method is similar to the method by Zingaretti et al. [248]. They also defined a new crossover and three mutation methods for operations on the quadtrees. Pignalberi et al. [193, 250] applied genetic algorithms to search a large solution space that requires a manageable amount of computation time. They used a fitness function that combines different measures to compare different parameters and they concentrated on range images, where the distance between the object and a sensor is employed to colour the pixels. This method can be applied for segmenting the 2D images as well as the out surfaces of 3D objects.

Chapter 3
Self-supervised Grey Level Image Segmentation Using an Optimised MUSIG (OptiMUSIG) Activation Function

3.1 Introduction

Different types of image segmentation methods, both supervised and unsupervised, as discussed in Chap. 2, have been applied over the years for the purpose of image segmentation and extraction. Among different types of image segmentation techniques, image segmentation using neural networks is the primary area of concern in this book as neural networks have the ability of approximation, adaptation and graceful degradation. Different neural network topologies are used to segment images on the basis of the image description through discrimination of the constituent regions or object specific features like intensity, texture, position, shape, size and homogeneity of image regions.

Most of the neural network architectures accounted so far, possess adequate knowledge in the field of segmentation of binary objects from a binary noisy/noise-free image. Hopfield network [57, 161, 251], Kohonen's self-organising map [39, 42, 43] and Kosko's bidirectional associative memory model [60] are some of the well-known neural network architectures in this regard. When applied to segmentation, the functionality of these networks is essentially based on the discrimination strategy of the pixel-intensity values of the test images. A pixel-based segmentation by self-organising feature map (SOFM) is reported in [252].

As reported in Chap. 2, the multilayer self-organising neural network (MLSONN) [3] architecture is a feedforward neural network architecture that uses image pixel neighbourhood information for the purpose of image segmentation. This network has been applied to extract binary objects from a noisy binary image by applying some fuzzy measures of the outputs in the output layer of the network architecture. The network uses the backpropagation algorithm [37, 47] to adjust the network weights with a view to arriving at a convergent stable solution. However, this network architecture is characterised by the generalised bilevel/bipolar sigmoidal activation function and operates in a self-supervised mode.

© Springer International Publishing AG 2016
S. De et al., *Hybrid Soft Computing for Multilevel Image and Data Segmentation*, Computational Intelligence Methods and Applications, DOI 10.1007/978-3-319-47524-0_3

Originally, the MLSONN architecture is unable to segment a multilevel image comprising different shades of grey as it is not capable of differentiating multilevel responses. The bipolar transfer characteristics is the main cause of this inherent limitation of the MLSONN architecture. The aforementioned shortcomings can be got rid of by a complicated enlarged network structure comprising several MLSONN architectures, one each for the different greyscale objects in the target segmented/extracted multilevel image. In this case, however, the increased space and time complexities for image segmentation procedure are sacrificed.

A functional modification has been rendered to rectify the drawback of the MLSONN [3] architecture. A multilevel sigmoidal (MUSIG) activation function is introduced by Bhattacharyya et al. [72, 174] for mapping multilevel input information into multiple scales of grey. The MUSIG activation function is a multilevel version of the standard sigmoidal function. This multilevel function thereby incorporates multiscaling capability in a single MLSONN architecture. The different transition levels of the MUSIG activation function are defined on the basis of the number of greyscale objects and the representative greyscale intensity levels. Thus, the MLSONN architecture guided by the MUSIG activation function is applied to segment the input images into different levels of grey. Though real-life images generally exhibit a varied amount of heterogeneity, this approach presumes that the information contained in the images are homogeneous in nature.

This chapter is targeted to address the limitation of the MUSIG activation function as it is applied to the MLSONN architecture. A functional modification of the transfer characteristics of the architecture through incorporation of the optimised class levels in the standard MUSIG activation function, is depicted in this chapter. The optimised class boundaries needed to segment grey level images into different classes are generated by a genetic algorithm. The resultant optimised class boundaries are used to design an optimised MUSIG (OptiMUSIG) [128, 134, 136, 172, 177] activation function for effecting multilevel image segmentation using a single MLSONN architecture.

The chapter commences with a discussion of the basic mathematical concepts and definitions to be applied throughout this chapter and the remaining chapters of the book. The network structure and operation of the MLSONN architecture used for the extraction of objects from a binary image are discussed in Sect. 3.3. This section also discusses the limitations of the MLSONN architecture as regards to the segmentation of multilevel images. Section 3.4 introduces a optimised multilevel sigmoidal (OptiMUSIG) activation function, which is able to induce the optimised class levels in the standard MUSIG activation function so as to address the limitation of the MUSIG activation function. The algorithm for the designing of the OptiMUSIG activation function is also presented in this section. An overview of four standard quantitative measures of the efficiency of segmentation is provided in Sect. 3.5. The principle of segmentation of multilevel greyscale images by the MLSONN architecture using the OptiMUSIG activation function is discussed in Sect. 3.6. Implementation results manifesting the application of the MLSONN architecture, characterised by the OptiMUSIG activation function, to segment the multilevel greyscale images is reported in Sect. 3.7. A comparative study of the results of segmentation of the test images by this approach vis-a-vis by the conventional MUSIG activation function and the standard FCM algorithm, is also illustrated in this section. An application of this approach is demonstrated using a synthetic and two real-life multilevel images, viz. the Lena and Baboon images and two multilevel biomedical images of brain, viz., a Brain Neuroanatomy (MRI) image [253] and a Anatomy of the brain (MRI) image [254]. ρ [72], F due to Liu and Yang [255] and F' and Q due to Borsotti et al. [256], are used to evaluate the segmentation efficiency of the said approach. Results of segmentation using the OptiMUSIG activation function-based method show better performance over the conventional MUSIG activation function employing heuristic class responses. The OptiMUSIG function is also found to outperform the standard FCM algorithm in this regard. Section 3.8 draws a line of conclusion to the chapter.

3.2 Mathematical Prerequisites

An overview of fuzzy c-means algorithm and the complexity of the genetic algorithm are discussed in this section.

3.2.1 Fuzzy c-Means

The fuzzy c-means (FCM) data clustering algorithm, introduced by Bezdek [199], assigns each data point to a cluster to a degree specified by its membership value. Each data point in the universe of data points is classified in a fuzzy c-partition and assigned a membership value. Hence, each data point possess a partial membership value in each and every cluster.

Let $X = \{x_1, x_2, \ldots, x_N\}$ where $x_i \in \Re^n$ represents a given set of feature data. FCM attempts to minimise the cost function [193, 199]

$$J(\mathbf{U}, \mathbf{M}) = \sum_{i=1}^{c} \sum_{j=1}^{N} (u_{ij})^m D_{ij} \tag{3.1}$$

where $M = \{m_1, \ldots, m_c\}$ is the cluster prototype (mean or centre) matrix. $U = (u_{ij})_{c \times N}$ is the fuzzy partition matrix and $u_{ij} \in [0, 1]$ is the membership coefficient of the jth data point in the ith cluster. The values of matrix U should satisfy the following conditions [193, 199]

$$u_{ij} = [0, 1], \forall i = 1, \ldots, N, \forall j = 1, \ldots, c \tag{3.2}$$

$$\sum_{j=1}^{c} u_{ij} = 1, \forall i = 1, \ldots, N \qquad (3.3)$$

The exponent $m \in [1, \infty)$ is the fuzzification parameter and usually set to 2. The most commonly used distance norm between a data point x_j and m_i is the Euclidean distance denoted as $D_{ij} = D(x_j, m_i)$.

Minimisation of the cost function $J(U, V)$ is a nonlinear optimization problem, which can be performed with the following iterative algorithm [193]:

1. Fix values for c, m and a small positive number ε and initialize the fuzzy partition matrix (**U**) randomly. Set step variable $t = 0$.
2. Calculate (at $t = 0$) or update (at $t > 0$) the membership matrix **U** by

$$u_{ij}^{(t+1)} = \frac{1}{\sum_{l=1}^{c} \left(\frac{D_{lj}}{D_{ij}}\right)^{\frac{1}{(1-m)}}} \qquad (3.4)$$

for $i = 1, \ldots, c$ and $j = 1, \ldots, N$.
3. Update the prototype matrix **M** by

$$m_i^{(t+1)} = \frac{\sum_{j=1}^{N} (u_{ij}^{(t+1)})^m x_j}{\sum_{j=1}^{N} (u_{ij}^{(t+1)})^m} \qquad (3.5)$$

for $i = 1, \ldots, c$
4. Repeat steps **2–3** until $\| \mathbf{M}^{(t+1)} - \mathbf{M}^{(t)} \| < \varepsilon$

The conventional FCM algorithm suffers from the presence of noise in images. This algorithm has no capability to incorporate any information about spatial context. Moreover, the fuzzy partitioning approach, applied in FCM, is not an optimised method.

3.2.2 Complexity Analysis of Genetic Algorithm

A detailed complexity analysis of the genetic algorithm is depicted in this section [193].

Considering an elitist selection strategy, let, p be the probability that the elitist buffer is refreshed and q is the probability that the buffer remains unchanged, i. e, $q = 1 - p$. The maximum number of generations below which the buffer refreshment would lead the evolution process to start afresh is denoted as M. In other words, M consecutive generations with no buffer refreshment means termination of the evolution process and reporting the buffer content as the possible solution. The number of generations before buffer refreshment in the ith evolution is denoted as X_i, where $i \geq 1$. Clearly, $1 \leq X_i \leq M, \forall i \geq 1$.

Now, X_i follows truncated geometric distribution. Moreover, $X_i, i \geq 1$ are independent and identically distributed random variables.

Now,

$$p_k = P[X_i = k] = Cq^{k-1}p, \ 1 \leq k \leq M \tag{3.6}$$

such that, $\sum_{k=1}^{M} p_k = 1$. Here, C is a constant. This implies that,

$$\sum_{k=1}^{M} Cq^{k-1}p = 1 \Leftrightarrow Cp \sum_{k=1}^{M} q^{k-1} = 1 \Leftrightarrow Cp\frac{(1-q^M)}{(1-q)} = 1$$
$$\Leftrightarrow C(1 - q^M) = 1 \Leftrightarrow C = \frac{1}{1-q^M} \tag{3.7}$$

Therefore,

$$p_k = P[X_i = k] = \frac{1}{1 - q^M} p.q^{k-1}, \ 1 \leq k \leq M, i \geq 1 \tag{3.8}$$

Now, we refer to 'Success' as the event that the evolutionary process terminates in an evolution after completing M generations within that evolution. Naturally, 'Failure' means that the evolutionary process does not terminate in an evolution before completion of M generations within that evolution.

Naturally,

$$P(Failure) = \sum_{r=1}^{(M-1)} p_r = P[X < M] = \sum_{r=1}^{(M-1)} \frac{1}{(1-q^M)} pq^{r-1} =$$
$$p\frac{(1-q^{(M-1)})}{(1-q)(1-q^M)} = (\frac{1-q^{(M-1)}}{1-q^M}) \tag{3.9}$$

$$\Rightarrow P(Success) = 1 - (\frac{1 - q^{(M-1)}}{1 - q^M}) \tag{3.10}$$

If it is assumed that N evolutions are required for termination of the evolution process then this means that there are $(N - 1)$ failures and subsequent success at the Nth evolution and $N \geq 1$. Let, Z is a random variable following a geometric distribution, which represents the number of evolutions required for termination.

Hence, the probability of termination of the evolution process after N evolutions is given by [193],

$$P(Z = N) = (\frac{1 - q^{(M-1)}}{1 - q^M})^{(N-1)} = (1 - (\frac{1 - q^{(M-1)}}{1 - q^M})), \ N \geq 1 \tag{3.11}$$

Since, Z follows geometric distribution, the expected convergence is given by,

$$E(Z) = \frac{1 - (\frac{1-q^{(M-1)}}{1-q^M})}{(\frac{1-q^{(M-1)}}{1-q^M})} = \frac{\frac{(1-q^M)-(1-q^{(M-1)})}{(1-q^M)}}{(\frac{1-q^{(M-1)}}{1-q^M})} = \frac{q^{(M-1)}(1-q)}{(1-q^{(M-1)})} = \frac{p(1-p)^{(M-1)}}{1 - (1-p)^{(M-1)}} \tag{3.12}$$

3.3   Multilayer Self-organising Neural Network (MLSONN) Architecture

A single multilayer self-organising neural network (MLSONN) architecture [3] is efficient in extracting binary objects from a noisy image. This feedforward neural network architecture operates in a self-supervised manner. It comprises an input layer, any number of hidden layers and an output layer of neurons. The neurons of each layer are connected with the neighbours of the corresponding neuron in the previous layer following a neighbourhood-based topology. Each output layer neuron is connected with the corresponding input layer neurons on a one-to-one basis. A schematic diagram of a three-layer version of the MLSONN architecture, employing a second-order neighbourhood topology-based interconnectivity, is shown in Fig. 3.1.

3.3.1   Operating Principle

The input layer of the network receives inputs and propagates the information to the following network layers. The jth hidden layer neuron receives the contributions (I_i) from the neighbouring ith neurons of the corresponding neuron in the input layer via the neighbourhood topology-based interconnection weights (w_{ij}) between the neurons. The jth hidden layer neuron receives total contribution I_j given by [3, 72]

$$I_j = \sum_i I_i w_{ji} \qquad (3.13)$$

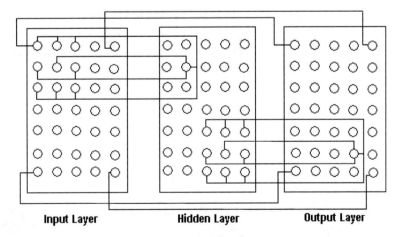

Fig. 3.1   A single three-layer self-organising neural network architecture with second-order neighbourhood topology-based connectivity

The hidden layer processes and propagates this incoming information to the output layer. The individual contribution H_j, from the jth hidden layer neuron towards an interconnected output layer neuron is decided as [3, 72]

$$H_j = f(I_j) \tag{3.14}$$

where, f is the standard bilevel sigmoidal activation function given by [3, 72]

$$y = f(x) = \frac{1}{1 + e^{-\lambda(x-\theta)}} \tag{3.15}$$

where λ decides the slope of the function and θ is a fixed threshold/bias value.

The network system errors at the output layer neurons are computed from the corresponding linear indices of fuzziness [3], which is specific to the network architecture. The interconnection weights are adjusted using the standard backpropagation algorithm. The next stage of processing is carried out by network architecture on the outputs fed back into the input layer from the output layer after the weights are adjusted. This processing is repeated until the interconnection weights stabilise or the system errors are reduced below some tolerable limits.

3.3.2 Network Error Adjustment

Basically, for a pattern p, the error(E) is measured as the difference between the desired value (T_i) and the output at any layer (O_i) and it is denoted as [3, 72]

$$E = \frac{1}{2} \sum_i (T_i - O_i)^2 \tag{3.16}$$

If the error measure E is a function of the linear index of fuzziness, the error adjustment term of the output layer of the network is obtained by taking the negative gradient of the error function as [3]

$$\Delta w_{ji} = \begin{cases} \psi(-\frac{2}{n}) O_{o,j}(1 - O_{o,j}) O_{h,i} & \text{if } \begin{matrix} 0 \leq O_{o,j} \leq 0.5 \\ 0.5 < O_{o,j} \leq 1 \end{matrix} \\ \psi(\frac{2}{n}) O_{o,j}(1 - O_{o,j}) O_{h,i} \end{cases} \tag{3.17}$$

where ψ is a constant, $O_{o,j}$ is the output of the jth neuron in the output layer and $O_{h,i}$ is the output of the ith neuron in the hidden layer.

The weight correction term in the other layers of the network is denoted as [3]

$$\Delta w_{ji} = \psi \left(\sum_k \delta_k w_{kj} \right) O_{o,j}(1 - O_{o,j}) O_{h,i} \tag{3.18}$$

where w_{kj} are the interconnection weights between the kth layer and jth layer neurons. δ_k is the negative gradient of the error function at the kth layer of the network. δ_k is denoted as [3, 72]

$$\delta_k = -\frac{\partial E}{\partial I_k} = -\frac{\partial E}{\partial O_k}\frac{\partial O_k}{\partial I_k} \tag{3.19}$$

where I_k is the contribution at the kth layer.

From Eq. 3.14,

$$\frac{\partial O_k}{\partial I_k} = f'(I_k) \tag{3.20}$$

Thus,

$$\delta_k = -\frac{\partial E}{\partial O_k}f'(I_k) \tag{3.21}$$

3.3.3 Self-Organisation Algorithm

A detailed algorithm of the operation of MLSONN architecture is presented in this section.

```
1 Begin
2 Read pix[1][m][n]
3 t:=0, wt[t][l][l+1]:=1, l=[1,2,3]
4 Do
5 Do
6 pix[l+1][m][n]=fsig[SUM[pix[l][m][n] x wt[t][l][l+1]]]
7 Loop for all layers
8 Do
9 Determine Deltawt[t][l][l+1] using backpropogation
   algorithm
10 Loop for all layers
11 pix[1][m][n]=pix[3][m][n]
12 Loop Until((wt[t][l][l-1]-wt[t-1][l][l-1])<eps)
13 End
```

Remark **pix[l][m][n]** are the fuzzified image pixel information at row m and column n at the network layers[l], i.e. the fuzzy membership values of the pixel intensities in the image scene. **pix[1][m][n]** are the fuzzy membership information of the input image scene and are fed as inputs to the input layer of the network. **pix[2][m][n]** and **pix[3][m][n]** are the corresponding information at the hidden and output layers. **fsig** is the standard sigmoidal activation function and **wt[t][l][l+1]** are the inter-layer interconnection weights between the network layers at a particular epoch t. **Deltawt[t][l][l+1]** is the weight correction term at a particular epoch. *eps* is the tolerable error.

The MLSONN architecture uses a sigmoidal activation function characterised by a bilevel response. Moreover, the architecture assumes homogeneity of the input image information. However, in real world situations, images exhibit a fair amount of heterogeneity in its information content which encompasses over the entire image pixel neighbourhoods. In the next section, we present two improved versions of the sigmoidal activation function to address these limitations.

3.4 Optimised Multilevel Sigmoidal (OptiMUSIG) Activation Function

It is evident that the standard bilevel sigmoidal activation function given in Eq. 3.15 produces bipolar responses [0(darkest)/1(brightest)] to incident input information. Hence, this activation function is unable to classify an image into multiple levels of grey. The bipolar form of the sigmoidal activation function has been extended into a multilevel sigmoidal (MUSIG) activation function [72] to generate multilevel outputs corresponding to input multilevel information. The multilevel form of the standard sigmoidal activation function is given by [72]

$$f_{\mathrm{MUSIG}}(x; \alpha_\gamma, c_\gamma) = \sum_{\gamma=1}^{K-1} \frac{1}{\alpha_\gamma + e^{-\lambda[x-(\gamma-1)c_\gamma-\theta]}} \qquad (3.22)$$

Figure 3.2a shows the conventional MUSIG [72] activation function generated for $K = 8$ with fixed and heuristic class responses.

In Eq. 3.22, α_γ represents the multilevel class responses given by [72]

$$\alpha_\gamma = \frac{C_N}{c_\gamma - c_{\gamma-1}} \qquad (3.23)$$

where γ represents the greyscale object index ($1 \leq \gamma < K$) and K is the number of classes of the segmented image. α_γ denotes the number of transition levels in the

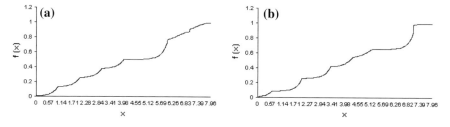

Fig. 3.2 a Conventional MUSIG activation function with fixed class responses **b** designed Opti-MUSIG activation function with optimised class responses

resulting MUSIG function. c_γ and $c_{\gamma-1}$ denote the greyscale contributions of the γth and $(\gamma-1)$th classes, respectively. C_N is the maximum fuzzy membership of the greyscale contribution of pixel neighbourhood geometry.

However, the class boundaries (c_γ) used by the MUSIG activation function are selected heuristically from the greyscale histograms of the input grey level images, assuming homogeneity of the underlying image information. Since real-life images are heterogeneous in nature, the class boundaries would differ from one image to another. So, optimised class boundaries derived from the image context would faithfully incorporate the intensity distribution of the images in the characteristic neuronal activations.

An optimised form of the MUSIG activation function, using optimised class boundaries can be represented as [128, 134, 136, 172, 177]

$$f_{\text{OptiMUSIG}} = \sum_{\gamma=1}^{K-1} \frac{1}{\alpha_{\gamma_{\text{opt}}} + e^{-\lambda[x-(\gamma-1)c_{\gamma_{\text{opt}}}-\theta]}} \tag{3.24}$$

where $c_{\gamma_{\text{opt}}}$ are the optimised greyscale contributions corresponding to optimised class boundaries. $\alpha_{\gamma_{\text{opt}}}$ are the respective optimised multilevel class responses. These parameters can be derived by suitable optimization of the segmentation of input images.

The algorithm for designing the OptiMUSIG activation function through the evolution of optimised class responses, is illustrated below. [193]

Algorithm for designing OptiMUSIG activation function

```
1 Begin
```

Generation of Optimised class boundaries

```
2 count:=0
3 Initialize Pop[count], K
4 Compute Fun(Pop[count])
5 Do
6 count:=count+1
7 Select Pop[count]
8 Crossover Pop[count]
9 Mutate Pop[count]
10 Loop Until (Fun(Pop[count])-Fun(Pop[count-1])<=eps)
11 clbound[K]:=Pop[count]
```

OptiMUSIG function generation phase

```
12 Generate OptiMUSIG with clbound[K]
13 End
```

Remark: **Pop[count]** is the initial population of class boundaries **clbound[K]** in the range [0, 255]. **K** is the number of target classes. **Fun** represents either of the fitness functions ρ, F, F' and Q. Selection of chromosomes based on better fitness value. GA crossover operation. GA mutation operation. eps is the tolerable error.

3.5 Evaluation of Segmentation Efficiency

Several unsupervised subjective measures have been proposed [257] to determine the segmentation efficiency of the existing segmentation algorithms. The following subsections discuss some of these measures. In this book, the segmentation policy is considered as follows: the K number of class levels are considered. Now, take a intensity level of a pixel of the original image. Find out that pixel intensity value nearer to which class level and replace the pixel intensity by that nearest class level in the segmented image. Repeat these steps for all the pixels in the image and the segmented image will be generated.

3.5.1 Correlation Coefficient (ρ)

The standard measure of correlation coefficient (ρ) [72] can be used to assess the quality of segmentation achieved. It is given by [72, 128, 134, 136, 172, 177]

$$\rho = \frac{\frac{1}{n^2}\sum_{i=1}^{n}\sum_{j=1}^{n}(I_{ij} - \overline{I})(S_{ij} - \overline{S})}{\sqrt{\frac{1}{n^2}\sum_{i=1}^{n}\sum_{j=1}^{n}(I_{ij} - \overline{I})^2}\sqrt{\frac{1}{n^2}\sum_{i=1}^{n}\sum_{j=1}^{n}(S_{ij} - \overline{S})^2}} \qquad (3.25)$$

where I_{ij}, $1 \leq i, j \leq n$ and S_{ij}, $1 \leq i, j \leq n$ are the original and the segmented images, respectively, each of dimensions $n \times n$. $\overline{I}$ and $\overline{S}$ are their respective mean intensity values. A higher value of ρ implies better quality of segmentation.

However, correlation coefficient has many limitations. The foremost disadvantage is that it is computationally intensive. This often confines its usefulness for image registration, i.e. orienting and positioning two images so they overlap. Moreover, the correlation coefficient is very much sensitive to image skewing, fading, etc. that inevitably occur in imaging systems.

3.5.2 Empirical Goodness Measures

In this subsection, an overview of three different empirical goodness measures is discussed.

Let S_I be the area of image (I) to be segmented into N number of regions. If R_k denotes the number of pixels in region k, then the area of region k is $S_k = |R_k|$. For the grey level intensity feature τ, let $C_\tau(p)$ denotes the value of τ for pixel p. The average value of τ in region k is then represented as [255]

$$\hat{C}_\tau(R_k) = \frac{\sum\limits_{p \varepsilon R_k} C_\tau(p)}{S_k} \tag{3.26}$$

The *squared colour error* of region k is then represented as [255]

$$e_k^2 = \sum_{\tau \varepsilon(r,g,b)} \sum_{p \varepsilon R_k} (C_\tau(p) - \hat{C}_\tau(R_k))^2 \tag{3.27}$$

Based on these notations three empirical measures (F, F' and Q) are described below.

3.5.2.1 Segmentation Efficiency Measure (F)

Liu and Yang proposed a quantitative evaluation function F for image segmentation [255]. It is given as

$$F(I) = \sqrt{N} \sum_{k=1}^{N} \frac{e_k^2}{\sqrt{S_k}} \tag{3.28}$$

3.5.2.2 Segmentation Efficiency Measure (F')

Borsotti et al. [256] proposed another evaluation function, F', to improve the performance of Liu and Yang's method [255]. The evaluation function, F', is represented as [255]

$$F'(I) = \frac{1}{S_I} \sqrt{\sum_{m=1}^{MaxArea} [N(m)]^{1+\frac{1}{m}}} \sum_{k=1}^{N} N \frac{e_k^2}{\sqrt{S_k}} \tag{3.29}$$

where $N(m)$ is represented as number of regions in the segmented image of an area of m and *MaxArea* is used as the area of the largest region in the segmented image.

3.5.2.3 Segmentation Efficiency Measure (Q)

Borsotti et al. [256] suggested another evaluation function, Q, to improve upon the performance of F and F'. The evaluation function, Q, is denoted as [256]

$$Q(I) = \frac{1}{1000.S_I} \sqrt{N} \sum_{k=1}^{N} \left[\frac{e_k^2}{1 + \log S_k} + \left(\frac{N(S_k)}{S_k} \right)^2 \right] \tag{3.30}$$

It may be noted that lower values of F, F' and Q imply better segmentation in contrast to the correlation coefficient (ρ) where higher values dictate terms.

3.6 Methodology

The approach of multilevel image segmentation by an OptiMUSIG activation function with a MLSONN architecture, has been implemented in three phases. The flow diagram is shown in Fig. 3.3. The different phases are discussed in the following subsections.

3.6.1 Generation of Optimised Class Boundaries

This is the most important phase of this approach. A genetic algorithm-based optimization procedure is used to generate the optimised class boundaries ($c_{\gamma_{opt}}$) of this OptiMUSIG activation function. The pixel intensity levels and the number of classes (K) are fed as inputs to a GA-based optimization procedure.

The genetic optimization procedure adopted for evolving optimised class boundaries from the input image information content uses a binary encoding technique for the chromosomes. A proportionate fitness selection operator is used to select the reproducing chromosomes, supported by a single point crossover operation. A population size of 50 has been used in this treatment.

Fig. 3.3 Flow diagram of image segmentation using OptiMUSIG activation function

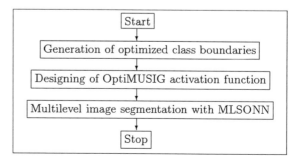

The segmentation efficiency measures (ρ, F, F', Q) given in Eqs. 3.25, 3.28, 3.29 and 3.30, respectively, are used as fitness functions for this phase. The selection probability of the ith chromosome is determined as [193]

$$p_i = \frac{f_i}{\sum\limits_{j=1}^{n} f_j} \qquad (3.31)$$

where f_i is the fitness value of the ith chromosome and n is the population size. The cumulative fitness P_i of each chromosome is evaluated by adding individual fitnesses in ascending order. Subsequently, the crossover and mutation operators are applied to evolve a new population.

3.6.2 Designing of OptiMUSIG Activation Function

In this phase, the optimised $c_{\gamma_{opt}}$ parameters obtained from the previous phase are used to determine the corresponding $\alpha_{\gamma_{opt}}$ parameters using Eq. 3.23. These $\alpha_{\gamma_{opt}}$ parameters are further employed to obtain the different transition levels of the OptiMUSIG activation function. Figure 3.2b shows a designed OptiMUSIG activation function for $K = 8$.

3.6.3 Multilevel Image Segmentation by OptiMUSIG

This is the final phase of this approach. A single MLSONN architecture guided by the designed OptiMUSIG activation function, is used to segment real-life multilevel images in this phase. The neurons of the different layers of the MLSONN architecture generate different grey level responses to the input image information. The processed input information propagates to the succeeding network layers. Since the network has no *a priori* knowledge about the output, the system errors are determined by the subnormal linear index of fuzziness (ν_{l_s}) as par Eq. 1.8 at the output layer of the MLSONN architecture. These errors are used to adjust the interconnection weights between the different layers using the standard backpropagation algorithm [37, 47]. The outputs at the output layer of the network is then fed back to the input layer for further processing to minimise the system errors. When the self-supervision of the network attains stabilisation, the original input image gets segmented into different multilevel regions depending upon the optimised transition levels of the OptiMUSIG activation function.

3.7 Results

The OptiMUSIG activation function-based approach has been applied for the segmentation of a synthetic multilevel image (Fig. 3.4 of dimensions 128×128) and two multilevel real-life images viz., Lena and Baboon (Figs. 3.5 and 3.6 of dimensions 128×128, 256×256 and 512×512). Experiments have been conducted with $K = \{6, 8\}$ classes [segmented outputs are reported for $K = 8$] for the previously mentioned test images. The OptiMUSIG activation function-based greyscale image segmentation is also demonstrated with two multilevel biomedical images of brain, viz., a Brain Neuroanatomy (MRI) image [253] and an Anatomy of the brain (MRI) image [254], each of dimensions 128×128 using the self-supervised MLSONN neural network architecture. The Brain Neuroanatomy (MRI) image and the Anatomy of the brain (MRI) image are shown in Figs. 3.7 and 3.8, respectively. The different components of the Brain Neuroanatomy (MRI) image shown in the Fig. 3.7 are: (i) Superior Frontal Gyrus (ii) Middle Frontal Gyrus (iii) Precentral Gyrus (iv) Inferior Temporal Gyrus (v) Cuneus (vi) Splenium of Corpus Callosum (vii) Skull (viii) Skin (ix) Fortix (x) Septum Pellucidum (xi) Genu of Corpus Callosum. In Fig. 3.8, the

Fig. 3.4 Original Synthetic image

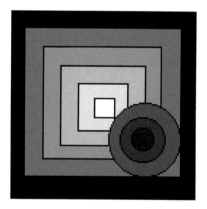

Fig. 3.5 Original Lena image

Fig. 3.6 Original Baboon image

Fig. 3.7 Brain Neuroanatomy (MRI) image with different components

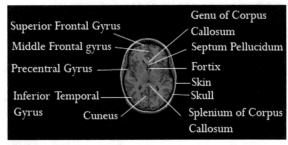

Fig. 3.8 Anatomy of the brain (MRI) image with different organs

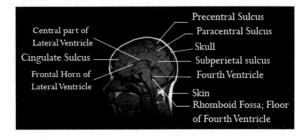

different organs depicted are: (i) Central part of Lateral Ventricle (ii) Cingulate Sulcus (iii) Frontal Horn of Lateral Ventricle (iv) Rhomboid Fossa; Floor of fourth Ventricle (v) Skin (vi) Fourth Ventricle (vii) Subperietal Sulcus (viii) Skull (ix) Paracentral Sulcus (x) Precentral Sulcus. We have only considered eight out of eleven different components of the first image for segmentation and detection. These are: (i) Middle Frontal Gyrus (ii) Precentral Gyrus (iii) Inferior Temporal Gyrus (iv) Cuneus (v) Splenium of Corpus Callosum (vi) Skull (vii) Skin (viii) Fortix. In the other MRI image, eight out of ten different organs are selected and they are: (i) Central part of Lateral Ventricle (ii) Cingulate Sulcus (iii) Frontal Horn of Lateral Ventricle (iv)

Rhomboid Fossa; Floor of fourth Ventricle (v) Skin (vi) Skull (vii) Paracentral Sulcus (viii) Precentral Sulcus. The experimental results and the segmented outputs of both the MRI images are reported for $K = 8$. The Brain Neuroanatomy (MRI) image and the Anatomy of the brain (MRI) image are noted in different tables and images as $MRI1$ and $MRI2$, respectively. The OptiMUSIG activation function has been designed with a fixed slope (λ) and a fixed threshold (θ) [results are reported for $\lambda = 4$ and $\theta = 2$]. Higher λ values lead to a faster convergence at the cost of segmentation quality and vice versa. Higher θ values lead to overthresholding, whereas lower values lead to underthresholding.

Sections 3.7.1.A and 3.7.2.A, respectively, discuss the segmentation efficiency of the OptiMUSIG activation function and the corresponding segmented outputs.

The presented approach has been compared with the segmentation achieved by means of the conventional MUSIG activation function with heuristic class levels and same number of classes. Furthermore, comparative studies with the standard FCM algorithm have also been carried out for the same number of classes.

Sections 3.7.1.B and 3.7.2.B, respectively, elaborate the performance of the conventional MUSIG activation function as regards to its efficacy in the segmentation of multilevel test images. The corresponding performance metrics and the segmented output images by the FCM algorithm are illustrated in Sects. 3.7.1.C and 3.7.2.C, respectively.

3.7.1 Quantitative Performance Analysis of Segmentation

This section illustrates the quantitative measures of evaluation of the segmentation efficiency of the OptiMUSIG, MUSIG and the FCM algorithm for $K = \{6, 8\}$ using the four measures of the segmentation efficiency viz., correlation coefficient (ρ) and evaluation functions (F, F' and Q). Section 3.7.1.A discusses the results obtained with the OptiMUSIG activation function. The corresponding results obtained with the conventional fixed class response-based MUSIG activation function and with the standard FCM algorithm, are provided in Sects. 3.7.1.B and 3.7.1.C, respectively.

3.7.1.1 OptiMUSIG-Guided Segmentation Evaluation

The optimised sets of class boundaries ($c_{\gamma_{opt}}$) obtained using genetic algorithm with four evaluation functions (ρ, F, F' and Q) and different number of classes are shown in Tables 3.1, 3.2, 3.3, 3.4, 3.5 and 3.6. The evaluation functions (EF), shown in the first columns of the tables, are applied as the fitness functions to generate genetic algorithm-based optimised class boundaries. All the evaluation function values are tabulated in normalised form in this chapter. The penultimate columns of the tables show the quality measures η [graded on a scale of **1** (best) to **4** (worst)] obtained by segmentation of the test images based on the corresponding set of optimised class boundaries. The best values obtained are indicated in the tables in **boldface** for easy

Table 3.1 Optimised class boundaries and evaluated segmentation quality measures for evaluation functions ρ, F, F' and Q for 6 classes for 128×128 sized image

EF	Image	Set	c_1	c_2	c_3	c_4	c_5	c_6	η	Avg. time
ρ	Lena	**1**	**0**	**84**	**115**	**143**	**182**	**255**	**0.8488 (1)**	8 m 56 s
		2	0	73	115	141	191	255	0.8347 (4)	
		3	0	83	114	142	181	255	0.8457 (2)	
		4	0	86	117	143	183	255	0.8404 (3)	
	Baboon	1	0	85	136	164	194	255	0.8647 (3)	4 m 55 s
		2	0	81	146	169	204	255	0.8721 (2)	
		3	0	79	148	168	201	255	0.8607 (4)	
		4	**0**	**81**	**143**	**165**	**202**	**255**	**0.8750 (1)**	
	Synthetic	1	0	61	100	131	174	255	0.8478 (3)	5 m 24 s
		2	0	57	99	129	179	255	0.8544 (2)	
		3	**0**	**49**	**104**	**132**	**184**	**255**	**0.8630 (1)**	
		4	0	51	99	132	184	255	0.8457 (4)	
F	Lena	1	0	100	105	154	160	255	0.1346 (2)	10 m 50 s
		2	0	97	106	161	164	255	0.1369 (4)	
		3	**0**	**100**	**106**	**157**	**165**	**255**	**0.1345 (1)**	
		4	0	95	106	153	163	255	0.1360 (3)	
	Baboon	1	0	100	101	133	144	255	0.1471 (3)	11 m 18 s
		2	**0**	**95**	**98**	**139**	**141**	**255**	**0.1459 (1)**	
		3	0	101	108	141	142	255	0.1471 (3)	
		4	0	98	101	137	145	255	0.1466 (2)	
	Synthetic	**1**	**0**	**56**	**85**	**133**	**137**	**255**	**0.1411 (1)**	10 m 55 s
		2	0	47	91	124	181	255	0.1414 (3)	
		3	0	55	88	133	163	255	0.1414 (3)	
		4	0	47	93	122	168	255	0.1413 (2)	
F'	Lena	1	0	99	101	151	164	255	0.1282 (4)	9 m 34 s
		2	**0**	**99**	**100**	**155**	**168**	**255**	**0.1278 (1)**	
		3	0	94	106	156	161	255	0.1279 (2)	
		4	0	99	105	154	168	255	0.1280 (3)	
	Baboon	1	0	98	102	136	140	255	0.1447 (2)	10 m 36 s
		2	0	104	105	142	146	255	0.1451 (3)	
		3	0	98	101	143	145	255	0.1465 (4)	
		4	**0**	**104**	**105**	**142**	**145**	**255**	**0.1434 (1)**	
	Synthetic	1	0	58	92	130	178	255	0.1418 (4)	10 m 05 s
		2	0	48	94	121	168	255	0.1417 (3)	
		3	0	50	85	132	163	255	0.1416 (2)	
		4	**0**	**59**	**80**	**137**	**156**	**255**	**0.1414 (1)**	

(continued)

Table 3.1 (continued)

EF	Image	Set	c_1	c_2	c_3	c_4	c_5	c_6	η	Avg. time
Q	Lena	**1**	**0**	**96**	**98**	**155**	**162**	**255**	**0.0731 (1)**	7 m 46 s
		2	0	97	101	146	153	255	0.0750 (4)	
		3	0	96	99	150	158	255	0.0735 (2)	
		4	0	93	102	159	163	255	0.0744 (3)	
	Baboon	1	0	106	107	137	145	255	0.1769 (3)	5 m 37 s
		2	**0**	**102**	**103**	**135**	**141**	**255**	**0.1722 (1)**	
		3	0	100	106	139	145	255	0.1781 (4)	
		4	0	96	99	137	1388	255	0.1741 (2)	
	Synthetic	**1**	**0**	**50**	**96**	**125**	**177**	**255**	**0.0673 (1)**	6 m 37 s
		2	**0**	**48**	**96**	**126**	**173**	**255**	**0.0673 (1)**	
		3	0	50	90	131	164	255	0.0674 (2)	
		4	0	47	99	137	170	255	0.0676 (3)	

Table 3.2 Optimised class boundaries and evaluated segmentation quality measures for evaluation functions ρ, F, F' and Q for 6 classes for 256×256 sized image

EF	Image	Set	c_1	c_2	c_3	c_4	c_5	c_6	η	Avg. time
ρ	Lena	**1**	**0**	**60**	**102**	**141**	**197**	**255**	**0.9570 (1)**	10 m 37 s
		2	0	69	115	151	190	255	0.9516 (2)	
		3	0	67	112	151	190	255	0.9469 (3)	
		4	0	60	102	142	192	255	0.9301 (4)	
	Baboon	**1**	**0**	**68**	**112**	**152**	**190**	**255**	**0.9320 (1)**	7 m 55 s
		2	0	67	107	147	185	255	0.9269 (2)	
		3	0	69	115	151	190	255	0.9088 (3)	
		4	0	67	112	151	190	255	0.9001 (4)	
F	Lena	1	0	77	91	159	166	255	0.2000 (2)	11 m 44 s
		2	**0**	**84**	**87**	**161**	**165**	**255**	**0.1995 (1)**	
		3	0	81	87	165	168	255	0.2002 (3)	
		4	0	71	99	159	164	255	0.2013 (4)	
	Baboon	1	0	98	99	149	150	255	0.3003 (2)	18 m 49 s
		2	**0**	**94**	**95**	**145**	**146**	**255**	**0.3000 (1)**	
		3	0	96	97	147	148	255	0.3004 (3)	
		4	0	97	99	149	150	255	0.3027 (4)	
F'	Lena	1	0	82	96	163	164	255	0.1807 (3)	11 m 44 s
		2	**0**	**77**	**90**	**163**	**169**	**255**	**0.1797 (1)**	
		3	0	72	94	162	164	255	0.1806 (2)	
		4	0	77	96	154	163	255	0.1808 (4)	
	Baboon	**1**	**0**	**92**	**93**	**143**	**144**	**255**	**0.2591 (1)**	10 m 28 s
		2	0	89	96	147	152	255	0.2637 (3)	
		3	0	95	98	149	153	255	0.2638 (4)	
		4	0	101	104	151	152	255	0.2635 (2)	

(continued)

Table 3.2 (continued)

EF	Image	Set	c_1	c_2	c_3	c_4	c_5	c_6	η	Avg. time
Q	Lena	**1**	**0**	**79**	**91**	**158**	**160**	**255**	**0.4277 (1)**	11 m 27 s
		2	0	81	89	154	157	255	0.4288 (3)	
		3	0	82	89	160	161	255	0.4280 (2)	
		4	0	80	89	159	162	255	0.4291 (4)	
	Baboon	**1**	**0**	**95**	**100**	**146**	**148**	**255**	**0.0769 (1)**	9 m 13 s
		2	0	95	99	151	152	255	0.0781 (4)	
		3	0	102	103	150	154	255	0.0779 (2)	
		4	0	102	103	143	145	255	0.0780 (3)	

Table 3.3 Optimised class boundaries and evaluated segmentation quality measures for evaluation functions ρ, F, F' and Q for 6 classes for 512×512 sized image

EF	Image	Set	c_1	c_2	c_3	c_4	c_5	c_6	η	Avg. time
ρ	Lena	1	0	49	107	148	190	255	0.9402	34 m 59 s
	Baboon	1	0	70	115	150	180	255	0.8922	45 m 01 s
F	Lena	1	0	107	114	160	164	255	0.2892	29 m 36 s
	Baboon	1	0	92	93	144	145	255	0.2671	30 m 15 s
F'	Lena	1	0	86	87	143	145	255	0.3074	28 m 31 s
	Baboon	1	0	90	91	141	143	255	0.3272	49 m 35 s
Q	Lena	1	0	107	114	160	162	255	0.1864	35 m 02 s
	Baboon	1	0	86	91	143	145	255	0.2840	29 m 26 s

Table 3.4 Optimised class boundaries and evaluated segmentation quality measures for evaluation functions ρ, F, F' and Q for 8 classes for 128×128 sized image

EF	Image	Set	c_1	c_2	c_3	c_4	c_5	c_6	c_7	c_8	η	Avg. time
ρ	Lena	1	0	46	96	131	155	171	207	255	0.9022 (3)	10 m 57 s
		2	**0**	**54**	**101**	**138**	**161**	**183**	**218**	**255**	**0.9123 (1)**	
		3	0	54	96	123	148	182	224	255	0.8757 (4)	
		4	0	46	96	126	146	177	206	255	0.9085 (2)	
	Baboon	1	0	58	86	123	141	156	194	255	0.8228 (2)	6 m 25 s
		2	0	67	104	124	140	160	189	255	0.8160 (3)	
		3	0	66	96	116	141	154	191	255	0.7763 (4)	
		4	**0**	**53**	**94**	**121**	**130**	**157**	**190**	**255**	**0.8258 (1)**	
	Synthetic	1	0	61	93	134	161	194	215	255	0.9168 (4)	8 m 31 s
		2	**0**	**52**	**98**	**130**	**160**	**209**	**229**	**255**	**0.9282 (1)**	
		3	0	52	91	137	165	205	221	255	0.9271 (2)	
		4	0	61	85	121	162	185	226	255	0.9207 (3)	

(continued)

Table 3.4 (continued)

EF	Image	Set	c_1	c_2	c_3	c_4	c_5	c_6	c_7	c_8	η	Avg. time
	$MRI1$	1	0	49	77	96	107	129	185	255	0.9832 (2)	7 m 23 s
		2	0	36	57	85	97	124	169	255	0.9855 (3)	
		3	0	42	65	87	108	130	239	255	0.9866 (4)	
		4	**0**	**36**	**66**	**93**	**120**	**134**	**197**	**255**	**0.9825 (1)**	
	$MRI2$	1	0	35	58	71	98	131	179	255	0.9238 (4)	6 m 56 s
		2	0	16	37	70	92	115	156	255	0.9302 (3)	
		3	**0**	**25**	**49**	**63**	**94**	**123**	**175**	**255**	**0.9360 (1)**	
		4	0	30	51	72	91	128	186	255	0.9359 (2)	
F	Lena	1	0	65	78	117	118	161	162	255	0.0785 (2)	11 m 27 s
		2	0	62	73	112	121	159	162	255	0.0812 (3)	
		3	**0**	**63**	**78**	**113**	**121**	**153**	**168**	**255**	**0.0783 (1)**	
		4	0	56	61	99	117	147	173	255	0.0819 (4)	
	Baboon	**1**	**0**	**73**	**79**	**115**	**117**	**149**	**151**	**255**	**0.1184 (1)**	8 m 07 s
		2	0	78	86	111	131	154	161	255	0.1207 (2)	
		3	0	67	98	111	133	151	153	255	0.1237 (4)	
		4	0	76	92	121	126	149	154	255	0.1218 (3)	
	Synthetic	1	0	69	70	102	120	159	161	255	0.0461 (4)	7 m 41 s
		2	**0**	**63**	**65**	**108**	**125**	**171**	**187**	**255**	**0.0449 (1)**	
		3	0	64	70	98	136	171	181	255	0.0451 (2)	
		4	0	64	74	97	140	167	185	255	0.0452 (3)	
	$MRI1$	1	0	42	58	71	88	111	122	255	0.1122 (3)	5 m 47 s
		2	**0**	**35**	**39**	**84**	**87**	**113**	**123**	**255**	**0.1103 (1)**	
		3	0	35	50	73	96	109	127	255	0.1129 (4)	
		4	0	33	47	64	98	108	123	255	0.1117 (2)	
	$MRI2$	1	0	36	58	82	85	154	169	255	0.1923 (3)	6 m 14 s
		2	0	34	50	66	96	137	175	255	0.1901 (2)	
		3	**0**	**29**	**42**	**75**	**84**	**155**	**158**	**255**	**0.1892 (1)**	
		4	0	27	32	71	81	157	164	255	0.1946 (4)	
F'	Lena	1	0	89	90	132	145	171	179	255	0.0723 (4)	5 m 29 s
		2	0	73	80	119	121	164	168	255	0.0723 (2)	
		3	**0**	**82**	**85**	**132**	**138**	**158**	**168**	**255**	**0.0719 (1)**	
		4	0	82	85	126	127	167	180	255	0.0723 (3)	
	Baboon	**1**	**0**	**86**	**90**	**114**	**132**	**147**	**157**	**255**	**0.1215 (1)**	11 m 39 s
		2	0	84	94	119	123	153	156	255	0.1230 (3)	
		3	0	70	73	118	124	150	151	255	0.1257 (4)	
		4	0	82	88	115	131	151	160	255	0.1219 (2)	
	Synthetic	1	0	60	73	102	132	150	191	255	0.0127 (3)	9 m 17 s
		2	**0**	**47**	**67**	**106**	**131**	**163**	**187**	**255**	**0.0123 (1)**	
		3	0	47	71	96	119	161	183	255	0.0124 (2)	
		4	0	47	65	104	118	168	184	255	0.0124 (2)	

(continued)

Table 3.4 (continued)

EF	Image	Set	c_1	c_2	c_3	c_4	c_5	c_6	c_7	c_8	η	Avg. time
	$MRI1$	1	0	43	50	76	87	117	121	255	0.0069 (1)	7 m 07 s
		2	0	30	43	78	85	113	121	255	0.0071 (3)	
		3	0	42	44	73	85	117	121	255	0.0070 (2)	
		4	0	31	44	68	93	110	127	255	0.0070 (2)	
	$MRI2$	1	0	21	35	68	88	146	172	255	0.0123 (2)	6 m 11 s
		2	**0**	**21**	**43**	**61**	**91**	**143**	**167**	**255**	**0.0122 (1)**	
		3	0	16	40	67	93	137	173	255	0.0123 (2)	
		4	0	20	36	71	93	143	161	255	0.0122 (1)	
Q	Lena	1	0	76	89	119	127	160	174	255	0.0136 (4)	6 m 04 s
		2	0	88	90	131	135	156	170	255	0.0134 (2)	
		3	**0**	**84**	**88**	**132**	**137**	**170**	**171**	**255**	**0.0129 (1)**	
		4	0	87	91	123	125	159	174	255	0.0135 (3)	
	Baboon	1	0	82	99	118	131	151	156	255	0.0426 (4)	6 m 24 s
		2	0	84	91	120	123	156	159	255	0.0418 (3)	
		3	**0**	**82**	**87**	**111**	**125**	**149**	**153**	**255**	**0.0404 (1)**	
		4	0	86	91	127	129	151	157	255	0.0414 (2)	
	Synthetic	1	0	49	62	109	119	174	191	255	0.0007 (2)	7 m 13 s
		2	**0**	**45**	**68**	**104**	**123**	**171**	**175**	**255**	**0.0006 (1)**	
		3	0	54	58	110	117	167	185	255	0.0007 (3)	
		4	0	51	64	108	124	167	186	255	0.0008 (4)	
	$MRI1$	**1**	**0**	**36**	**44**	**81**	**86**	**116**	**120**	**255**	**0.0048 (1)**	6 m 12 s
		2	0	18	52	73	90	115	123	255	0.0049 (2)	
		3	0	27	54	74	95	112	127	255	0.0049 (2)	
		4	**0**	**38**	**48**	**78**	**85**	**117**	**119**	**255**	**0.0048 (1)**	
	$MRI2$	1	0	25	47	64	98	129	183	255	0.0030 (2)	6 m 43 s
		2	0	30	36	74	78	142	160	255	0.0030 (2)	
		3	0	23	43	73	92	136	177	255	0.0031 (3)	
		4	**0**	**27**	**44**	**74**	**87**	**141**	**161**	**255**	**0.0029 (1)**	

reckoning. The average computation times ($Avg.Time$) are also shown in the table alongside.

3.7.1.2   MUSIG-Guided Segmentation Evaluation

Tables 3.7, 3.8, 3.9, 3.10, 3.11 and 3.12 show the heuristically selected class boundaries for the conventionalMUSIG activation function used in the segmentation of the test images along with the corresponding quality measure values evaluated after the segmentation process. Similar to the optimised results, the best values obtained are also indicated in these tables in **boldface**.

Table 3.5 Optimised class boundaries and evaluated segmentation quality measures for evaluation functions ρ, F, F' and Q for 8 classes for 256×256 sized images

EF	Image	Set	c_1	c_2	c_3	c_4	c_5	c_6	c_7	c_8	η	Avg. time
ρ	Lena	1	0	21	65	106	138	167	227	255	0.9620 (4)	11 m 39 s
		2	**0**	**41**	**80**	**123**	**158**	**201**	**240**	**255**	**0.9718 (1)**	
		3	0	38	71	113	135	168	209	255	0.9681 (2)	
		4	0	47	80	111	146	168	212	255	0.9673 (3)	
	Baboon	1	0	68	91	112	141	169	190	255	0.9112 (3)	12 m 21 s
		2	**0**	**66**	**84**	**116**	**143**	**166**	**189**	**255**	**0.9235 (1)**	
		3	0	65	104	128	152	168	191	255	0.9111 (4)	
		4	0	40	78	114	140	164	191	255	0.9143 (2)	
F	Lena	**1**	**0**	**47**	**61**	**111**	**120**	**169**	**172**	**255**	**0.1072 (1)**	24 m 41 s
		2	0	48	55	121	135	185	195	255	0.1101 (4)	
		3	0	47	53	117	119	167	172	255	0.1074 (2)	
		4	0	53	55	119	121	170	186	255	0.1095 (3)	
	Baboon	1	0	82	90	125	133	163	169	255	0.1672 (2)	10 m 02 s
		2	**0**	**81**	**82**	**124**	**126**	**159**	**162**	**255**	**0.1652 (1)**	
		3	0	76	82	108	114	147	150	255	0.1678 (4)	
		4	0	85	86	125	132	164	173	255	0.1674 (3)	
F'	Lena	**1**	**0**	**48**	**69**	**115**	**121**	**164**	**175**	**255**	**0.0940 (1)**	9 m 53 s
		2	0	52	62	113	139	183	197	255	0.0949 (3)	
		3	0	51	66	120	128	181	186	255	0.0945 (2)	
		4	0	48	59	120	123	175	182	255	0.1087 (4)	
	Baboon	**1**	**0**	**82**	**83**	**114**	**121**	**152**	**154**	**255**	**0.1640 (1)**	12 m 20 s
		2	0	78	82	125	126	162	165	255	0.1652 (2)	
		3	0	82	85	121	132	163	165	255	0.1659 (3)	
		4	0	74	89	121	136	165	166	255	0.1683 (4)	
Q	Lena	1	0	46	53	117	120	170	179	255	0.0024 (4)	8 m 43 s
		2	0	44	59	111	122	163	171	255	0.0022 (2)	
		3	**0**	**50**	**59**	**119**	**120**	**170**	**171**	**255**	**0.0021 (1)**	
		4	0	53	68	112	115	168	174	255	0.0023 (3)	
	Baboon	**1**	**0**	**81**	**85**	**119**	**124**	**152**	**153**	**255**	**0.0010 (1)**	13 m 06 s
		2	0	79	91	127	129	155	160	255	0.0012 (3)	
		3	0	82	88	124	131	155	157	255	0.0011 (2)	
		4	0	74	79	121	124	154	156	255	0.0013 (4)	

Table 3.6 Optimised class boundaries and evaluated segmentation quality measures for evaluation functions ρ, F, F' and Q for 8 classes for 512×512 sized images

EF	Image	Set	c_1	c_2	c_3	c_4	c_5	c_6	c_7	c_8	η	Avg. time
ρ	Lena	1	0	40	80	110	150	180	210	255	0.9644	37 m 17 s
	Baboon	1	0	63	94	125	142	169	191	255	0.9412	37 m 47 s
F	Lena	1	0	85	89	134	146	181	184	255	0.2101	41 m 53 s
	Baboon	1	0	69	75	116	121	151	155	255	0.1720	50 m 03 s
F'	Lena	1	0	86	91	140	141	185	190	255	0.2135	31 m 36 s
	Baboon	1	0	78	81	121	132	163	172	255	0.1732	57 m 08 s
Q	Lena	1	0	88	93	135	142	180	189	255	0.0297	47 m 16 s
	Baboon	1	0	85	94	129	131	171	172	255	0.0067	38 m 16 s

Table 3.7 Fixed class boundaries and evaluated segmentation quality measures for evaluation functions ρ, F, F' and Q for 6 classes for 128×128 sized images

EF	Image	Set	c_1	c_2	c_3	c_4	c_5	c_6	η
ρ	Lena	1	0	75	114	140	198	255	0.8039 (2)
		2	**0**	**76**	**110**	**142**	**194**	**255**	**0.8076 (1)**
		3	0	80	120	150	220	255	0.7768 (4)
		4	0	80	125	145	215	255	0.7826 (3)
	Baboon	1	0	70	106	134	174	255	0.7463 (3)
		2	0	70	97	132	174	255	0.5935 (4)
		3	0	71	103	128	165	255	0.7563 (2)
		4	**0**	**72**	**115**	**147**	**179**	**255**	**0.8214 (1)**
	Synthetic	1	0	47	98	129	184	255	0.8049 (2)
		2	0	55	93	141	177	255	0.6818 (4)
		3	0	52	99	135	172	255	0.7409 (3)
		4	**0**	**61**	**97**	**132**	**182**	**255**	**0.8225 (1)**
F	Lena	1	0	90	125	145	190	255	0.1655 (3)
		2	**0**	**90**	**100**	**140**	**180**	**255**	**0.1512 (1)**
		3	0	85	105	145	175	255	0.1532 (2)
		4	0	80	115	140	170	255	0.1532 (2)
	Baboon	1	0	105	115	145	150	255	0.1609 (4)
		2	0	102	112	142	152	255	0.1592 (3)
		3	**0**	**95**	**110**	**135**	**145**	**255**	**0.1554 (1)**
		4	0	98	108	133	148	255	0.1560 (2)
	Synthetic	**1**	**0**	**50**	**90**	**140**	**160**	**255**	**0.1417 (1)**
		2	0	50	90	140	170	255	0.1419 (2)
		3	0	58	92	142	172	255	0.1465 (4)
		4	0	45	88	125	175	255	0.1445 (3)

(continued)

Table 3.7 (continued)

EF	Image	Set	c_1	c_2	c_3	c_4	c_5	c_6	η
F'	Lena	1	0	85	95	155	165	255	0.1399 (2)
		2	**0**	**90**	**96**	**145**	**166**	**255**	**0.1374 (1)**
		3	0	94	106	140	156	255	0.1452 (3)
		4	0	98	108	138	156	255	0.1501 (4)
	Baboon	1	0	88	95	135	165	255	0.1945 (4)
		2	0	95	115	135	165	255	0.1808 (3)
		3	**0**	**96**	**105**	**134**	**155**	**255**	**0.1620 (1)**
		4	0	90	100	130	150	255	0.1627 (2)
	Synthetic	1	0	55	95	135	190	255	0.1604 (3)
		2	**0**	**50**	**90**	**115**	**165**	**255**	**0.1434 (1)**
		3	0	45	90	135	155	255	0.1437 (2)
		4	0	54	75	132	151	255	0.1437 (2)
Q	Lena	1	0	80	95	145	165	255	0.0849 (4)
		2	0	85	95	142	160	255	0.0815 (3)
		3	0	86	96	144	162	255	0.0806 (2)
		4	**0**	**90**	**97**	**148**	**165**	**255**	**0.0773 (1)**
	Baboon	1	0	105	110	130	155	255	0.2427 (4)
		2	0	100	115	135	150	255	0.2098 (2)
		3	0	98	118	138	158	255	0.2233 (3)
		4	**0**	**102**	**114**	**134**	**154**	**255**	**0.2093 (1)**
	Synthetic	1	0	55	90	120	160	255	0.0690 (3)
		2	0	50	88	123	175	255	0.0684 (2)
		3	0	45	85	126	159	255	0.0693 (4)
		4	**0**	**42**	**94**	**132**	**165**	**255**	**0.0680 (1)**

Table 3.8 Fixed class boundaries and evaluated segmentation quality measures for evaluation functions ρ, F, F' and Q for 6 classes for 256×256 sized images

EF	Image	Set	c_1	c_2	c_3	c_4	c_5	c_6	η
ρ	Lena	1	0	50	100	125	150	255	0.9026 (2)
		2	**0**	**40**	**90**	**125**	**205**	**255**	**0.9067 (1)**
		3	0	50	85	135	215	255	0.8967 (3)
		4	0	35	105	165	210	255	0.8700 (4)
	Baboon	1	0	35	74	120	150	255	0.8601 (2)
		2	0	50	100	120	150	255	0.8223 (4)
		3	**0**	**50**	**100**	**150**	**175**	**255**	**0.8837 (1)**
		4	0	60	105	145	200	255	0.8413 (3)

(continued)

Table 3.8 (continued)

EF	Image	Set	c_1	c_2	c_3	c_4	c_5	c_6	η
F	Lena	1	0	40	65	85	115	255	0.3175 (4)
		2	0	70	110	145	205	255	0.2285 (2)
		3	**0**	**80**	**100**	**140**	**180**	**255**	**0.2172 (1)**
		4	0	90	125	145	190	255	0.2545 (3)
	Baboon	1	0	85	95	150	185	255	0.4234 (2)
		2	0	90	100	145	200	255	0.4541 (4)
		3	**0**	**100**	**115**	**155**	**180**	**255**	**0.4065 (1)**
		4	0	105	125	165	190	255	0.4369 (3)
F'	Lena	**1**	**0**	**70**	**90**	**160**	**190**	**255**	**0.1944 (1)**
		2	0	65	85	150	185	255	0.1984 (3)
		3	0	55	95	160	200	255	0.2164 (4)
		4	0	85	95	155	195	255	0.1966 (2)
	Baboon	1	0	60	85	125	175	255	0.4216 (4)
		2	0	75	85	135	165	255	0.34525 (3)
		3	0	80	90	130	160	255	0.3263 (2)
		4	**0**	**85**	**95**	**135**	**165**	**255**	**0.3184 (1)**
Q	Lena	**1**	**0**	**80**	**90**	**150**	**160**	**255**	**0.3590 (1)**
		2	0	85	95	155	165	255	0.3592 (2)
		3	0	95	100	165	175	255	0.3963 (4)
		4	0	80	95	145	165	255	0.3685 (3)
	Baboon	1	0	85	95	150	165	255	0.0994 (3)
		2	**0**	**95**	**105**	**145**	**160**	**255**	**0.0860 (1)**
		3	0	95	100	135	155	255	0.0910 (2)
		4	0	105	110	130	160	255	0.1064 (4)

Table 3.9 Fixed class boundaries and evaluated segmentation quality measures for evaluation functions ρ, F, F' and Q for 6 classes for 512×512 sized images

EF	Image	Set	c_1	c_2	c_3	c_4	c_5	c_6	η
ρ	Lena	1	0	40	100	155	190	255	0.9024
	Baboon	1	0	67	107	141	188	255	0.8123
F	Lena	1	0	80	90	135	150	255	0.3326
	Baboon	1	0	80	95	135	155	255	0.3161
F'	Lena	1	0	110	120	160	175	255	0.3235
	Baboon	1	0	80	90	130	145	255	0.3502
Q	Lena	1	0	105	110	160	175	255	0.1996
	Baboon	1	0	90	105	140	155	255	0.3128

Table 3.10 Fixed class boundaries and evaluated segmentation quality measures for evaluation functions ρ, F, F' and Q for 8 classes for 128×128 sized images

EF	Image	Set	c_1	c_2	c_3	c_4	c_5	c_6	c_7	c_8	η
ρ	Lena	1	0	40	90	120	145	165	210	255	0.8880 (3)
		2	**0**	**45**	**91**	**128**	**150**	**170**	**218**	**255**	**0.9001 (1)**
		3	0	64	100	128	149	185	238	255	0.8645 (4)
		4	0	46	96	130	140	187	219	255	0.8968 (2)
	Baboon	**1**	**0**	**50**	**80**	**120**	**140**	**150**	**190**	**255**	**0.8183 (1)**
		2	0	65	100	120	130	150	180	255	0.8037 (2)
		3	0	65	87	110	131	167	200	255	0.7631 (4)
		4	0	59	84	111	140	165	195	255	0.6801(3)
	Synthetic	1	0	50	96	129	157	198	240	255	0.8179 (4)
		2	**0**	**43**	**92**	**127**	**162**	**192**	**214**	**255**	**0.9062 (1)**
		3	0	54	77	102	135	155	186	255	0.8918 (2)
		4	0	41	90	124	153	181	229	255	0.8650 (3)
	$MRI1$	1	0	25	60	90	105	120	170	255	0.9638 (3)
		2	0	25	60	90	100	120	175	255	0.9245 (4)
		3	**0**	**30**	**60**	**85**	**100**	**120**	**175**	**255**	**0.9783 (1)**
		4	0	30	65	90	110	125	175	255	0.9669 (2)
	$MRI2$	1	0	30	55	75	90	130	180	255	0.8840 (4)
		2	0	35	65	80	90	125	180	255	0.9134 (2)
		3	**0**	**38**	**60**	**78**	**90**	**125**	**185**	**255**	**0.9144 (1)**
		4	0	35	75	85	100	130	185	255	0.9043 (3)
F	Lena	1	0	40	65	85	115	145	175	255	0.0921 (2)
		2	**0**	**60**	**70**	**110**	**115**	**140**	**170**	**255**	**0.0861 (1)**
		3	0	50	77	100	120	140	160	255	0.0957 (3)
		4	0	45	65	100	138	155	180	255	0.1073 (4)
	Baboon	1	0	70	77	113	119	159	161	255	0.1284 (2)
		2	0	75	84	113	129	155	171	255	0.1321 (3)
		3	0	65	94	123	139	155	166	255	0.1415 (4)
		4	**0**	**70**	**90**	**120**	**130**	**150**	**170**	**255**	**0.0803 (1)**
	Synthetic	1	0	60	75	110	120	150	165	255	0.1150 (3)
		2	**0**	**60**	**70**	**110**	**120**	**160**	**185**	**255**	**0.0490 (1)**
		3	0	65	75	105	120	150	165	255	0.1102 (3)
		4	0	65	75	105	125	145	165	255	0.1074 (2)
	$MRI1$	**1**	**0**	**40**	**60**	**75**	**90**	**110**	**130**	**255**	**0.1180 (1)**
		2	0	35	55	70	90	105	135	255	0.1279 (2)
		3	0	30	60	70	85	115	135	255	0.1438 (3)
		4	0	38	60	72	88	108	140	255	0.1517 (4)
	$MRI2$	**1**	**0**	**40**	**60**	**85**	**100**	**145**	**170**	**255**	**0.1996 (1)**
		2	0	35	65	85	90	140	160	255	0.2005 (2)
		3	0	35	55	90	105	145	165	255	0.2039 (3)
		4	0	42	65	93	105	142	165	255	0.2097 (4)

(continued)

Table 3.10 (continued)

EF	Image	Set	c_1	c_2	c_3	c_4	c_5	c_6	c_7	c_8	η
F'	Lena	1	0	74	85	115	125	145	160	255	0.0818 (4)
		2	0	78	92	130	140	155	165	255	0.0773 (3)
		3	0	80	95	120	135	155	165	255	0.0771 (2)
		4	**0**	**80**	**88**	**115**	**132**	**158**	**166**	**255**	**0.0729 (1)**
	Baboon	1	0	79	91	124	138	165	178	255	0.1374 (3)
		2	0	80	90	120	130	160	180	255	0.1356 (2)
		3	0	85	90	110	135	150	170	255	0.1390 (4)
		4	**0**	**75**	**85**	**115**	**125**	**145**	**160**	**255**	**0.1254 (1)**
	Synthetic	**1**	**0**	**70**	**80**	**100**	**130**	**150**	**180**	**255**	**0.0146 (1)**
		2	0	50	70	110	130	160	190	255	0.0151 (2)
		3	0	44	75	105	125	165	195	255	0.0196 (3)
		4	0	44	75	109	125	165	195	255	0.0198 (4)
	$MRI1$	1	0	35	55	75	100	120	135	255	0.0087 (3)
		2	0	25	50	80	100	115	130	255	0.0075 (2)
		3	0	25	55	75	90	110	130	255	0.0075 (2)
		4	**0**	**30**	**50**	**80**	**90**	**115**	**130**	**255**	**0.0074 (1)**
	$MRI2$	1	0	33	66	88	111	133	166	255	0.0148 (4)
		2	0	20	55	85	95	135	165	255	0.0132 (2)
		3	**0**	**25**	**40**	**75**	**95**	**130**	**160**	**255**	**0.0131 (1)**
		4	0	30	50	80	100	125	160	255	0.0136 (3)
Q	Lena	1	0	74	80	125	132	151	165	255	0.0155 (3)
		2	0	70	85	120	135	150	165	255	0.0161 (4)
		3	**0**	**78**	**88**	**128**	**139**	**156**	**168**	**255**	**0.0141 (1)**
		4	0	81	90	128	136	153	170	255	0.0142 (2)
	Baboon	1	0	70	85	115	135	150	175	255	0.0501 (3)
		2	**0**	**75**	**88**	**110**	**130**	**145**	**165**	**255**	**0.0467 (1)**
		3	0	80	90	115	132	147	175	255	0.0512 (4)
		4	0	74	80	125	132	151	165	255	0.0476 (2)
	Synthetic	1	0	50	60	100	120	170	180	255	0.0006 (2)
		2	**0**	**55**	**70**	**105**	**130**	**165**	**192**	**255**	**0.0001 (1)**
		3	0	50	65	115	130	160	185	255	0.0007 (3)
		4	0	55	70	105	130	165	192	255	0.0011 (4)
	$MRI1$	1	0	25	50	75	100	115	135	255	0.0056 (3)
		2	0	30	60	75	90	115	130	255	0.0055 (2)
		3	0	28	58	78	98	118	138	255	0.0061 (4)
		4	**0**	**33**	**63**	**83**	**95**	**113**	**133**	**255**	**0.0054 (1)**
	$MRI2$	1	0	35	60	70	85	115	175	255	0.0037 (4)
		2	0	35	60	75	90	125	160	255	0.0035 (3)
		3	**0**	**30**	**50**	**70**	**100**	**120**	**165**	**255**	**0.0031 (1)**
		4	0	25	50	75	100	125	175	255	0.0032 (2)

Table 3.11 Fixed class boundaries and evaluated segmentation quality measures for evaluation functions ρ, F, F' and Q for 8 classes for 256×256 sized images

EF	Image	Set	c_1	c_2	c_3	c_4	c_5	c_6	c_7	c_8	η
ρ	Lena	1	0	32	64	96	128	192	224	255	0.8927 (2)
		2	**0**	**38**	**82**	**115**	**145**	**175**	**213**	**255**	**0.8952 (1)**
		3	0	34	72	105	128	155	193	255	0.8903 (3)
		4	0	53	78	111	148	187	224	255	0.8675 (4)
	Baboon	1	0	30	60	90	120	160	200	255	0.8640 (3)
		2	0	25	95	110	140	170	210	255	0.8792 (2)
		3	**0**	**30**	**90**	**115**	**145**	**175**	**215**	**255**	**0.8826 (1)**
		4	0	25	95	120	140	175	215	255	0.8033 (4)
F	Lena	1	0	30	60	90	120	150	180	255	0.1218 (2)
		2	0	40	67	110	125	140	190	255	0.1274 (4)
		3	**0**	**35**	**55**	**100**	**138**	**155**	**180**	**255**	**0.1206 (1)**
		4	0	45	58	110	128	145	200	255	0.1264 (3)
	Baboon	**1**	**0**	**70**	**90**	**120**	**140**	**160**	**180**	**255**	**0.1794 (1)**
		2	0	65	85	115	135	155	185	255	0.1876 (4)
		3	0	68	88	118	138	158	188	255	0.1859 (3)
		4	0	73	85	113	125	143	160	255	0.1806 (2)
F'	Lena	1	0	40	70	110	120	160	200	255	0.1044 (3)
		2	0	35	65	105	115	165	205	255	0.1078 (4)
		3	0	45	75	105	125	155	195	255	0.1027 (2)
		4	**0**	**43**	**73**	**113**	**133**	**153**	**193**	**255**	**0.1016 (1)**
	Baboon	1	0	80	90	125	135	170	190	255	0.1926 (4)
		2	0	80	95	120	136	160	190	255	0.1866 (3)
		3	0	85	93	122	133	165	188	255	0.1831 (2)
		4	**0**	**79**	**91**	**124**	**138**	**165**	**178**	**255**	**0.1747 (1)**
Q	Lena	1	0	38	63	105	125	155	205	255	0.0068 (4)
		2	0	40	65	85	115	145	175	255	0.0067 (3)
		3	**0**	**55**	**75**	**120**	**130**	**165**	**190**	**255**	**0.0064 (1)**
		4	0	40	60	100	130	165	190	255	0.0065 (2)
	Baboon	1	0	80	90	120	130	145	175	255	0.0014 (4)
		2	**0**	**82**	**92**	**122**	**132**	**150**	**165**	**255**	**0.0011 (1)**
		3	0	75	90	125	140	155	180	255	0.0012 (2)
		4	0	70	85	115	135	150	175	255	0.0013 (3)

3.7.1.3 FCM Guided Segmentation Evaluation

The standard FCM algorithm has been applied for the segmentation of the test multilevel images. As already stated, the FCM algorithm resorts to an initial random selection of the cluster centroids out of the image data. The algorithm converges to a known number of cluster centroids. Tables 3.13, 3.14, 3.15, 3.16, 3.17 and 3.18

Table 3.12 Fixed class boundaries and evaluated segmentation quality measures for evaluation functions ρ, F, F' and Q for 8 classes for 256×256 sized images

EF	Image	Set	c_1	c_2	c_3	c_4	c_5	c_6	c_7	c_8	η
ρ	Lena	1	0	40	80	110	130	190	224	255	0.9363
	Baboon	1	0	38	81	118	140	165	195	255	0.9057
F	Lena	1	0	75	95	120	140	170	180	255	0.2389
	Baboon	1	0	75	80	120	140	150	185	255	0.1882
F'	Lena	1	0	80	90	130	160	180	200	255	0.2446
	Baboon	1	0	70	85	120	140	155	180	255	0.1829
Q	Lena	1	0	80	95	120	135	160	180	255	0.0333
	Baboon	1	0	90	100	135	145	175	195	255	0.0075

Table 3.13 FCM guided class boundaries and evaluated segmentation quality measures for evaluation functions ρ, F, F' and Q for 6 classes for 128×128 sized images

EF	Image	Set	c_1	c_2	c_3	c_4	c_5	c_6	η
ρ	Lena	1	80	130	53	157	103	196	0.8409 (2)
		2	55	140	162	198	118	91	**0.8434 (1)**
	Baboon	1	66	135	115	157	94	182	0.8666 (2)
		2	136	182	115	66	157	94	**0.8667 (1)**
	Synthetic	1	0	79	48	100	197	133	**0.9434 (1)**
		2	48	0	79	100	197	133	**0.9434 (1)**
F	Lena	1	146	167	122	93	200	55	**1.0000 (2)**
		2	200	93	168	122	146	55	0.9996 (1)
	Baboon	1	93	66	115	135	156	182	0.9999 (1)
		2	156	182	94	115	135	66	**1.0000 (2)**
	Synthetic	1	48	133	0	100	79	197	**1.0000 (1)**
		2	133	100	0	48	79	197	**1.0000 (1)**
F'	Lena	1	86	133	54	158	109	197	**1.0000 (2)**
		2	198	160	137	114	89	54	0.9926 (1)
	Baboon	1	115	136	157	182	94	66	0.9959 (1)
		2	94	182	135	66	115	156	**1.0000 (2)**
	Synthetic	1	100	0	79	133	48	197	**1.0000 (1)**
		2	197	100	48	0	79	133	**1.0000 (1)**
Q	Lena	1	146	55	93	122	200	168	**1.0000 (2)**
		2	200	145	167	121	93	55	0.9274 (1)
	Baboon	1	115	157	136	94	66	182	**1.0000 (2)**
		2	94	115	135	182	66	156	0.6817 (1)
	Synthetic	1	53	93	160	203	127	0	**1.0000 (1)**
		2	160	93	127	203	53	0	**1.0000 (1)**

Table 3.14 FCM guided class boundaries and evaluated segmentation quality measures for evaluation functions ρ, F, F' and Q for 6 classes for 256×256 sized images

EF	Image	Set	c_1	c_2	c_3	c_4	c_5	c_6	η
ρ	Lena	**1**	**63**	**10**	**175**	223	140	105	**0.9123 (1)**
		2	224	176	141	106	63	10	0.9120 (2)
	Baboon	1	65	95	159	138	184	117	0.8979 (2)
		2	**65**	**95**	**138**	**117**	**159**	**184**	**0.8978 (1)**
F	Lena	1	170	10	103	61	136	222	1.0000 (2)
		2	**10**	**62**	**104**	**173**	**223**	**139**	**0.9417 (1)**
	Baboon	**1**	**65**	**159**	**138**	**117**	**184**	**95**	**0.9647 (1)**
		2	184	138	117	95	159	65	1.0000 (2)
F'	Lena	1	176	106	10	63	224	141	1.0000 (2)
		2	**10**	**63**	**106**	**224**	**141**	**176**	**0.9999 (1)**
	Baboon	**1**	**117**	**159**	**95**	**138**	**184**	**65**	**0.9999 (1)**
		2	95	138	117	159	184	65	1.0000 (2)
Q	Lena	**1**	**63**	**141**	**10**	**224**	**176**	**106**	**0.7088 (1)**
		2	174	223	63	140	10	105	1.0000 (2)
	Baboon	**1**	**138**	**65**	**95**	**184**	**117**	**159**	**0.9999 (1)**
		2	95	138	184	65	117	159	1.0000 (2)

Table 3.15 FCM guided class boundaries and evaluated segmentation quality measures for evaluation functions ρ, F, F' and Q for 6 classes for 512×512 sized images

EF	Image	Set	c_1	c_2	c_3	c_4	c_5	c_6	η
ρ	Lena	1	133	160	79	105	201	50	0.9437
	Baboon	1	115	158	89	182	135	56	0.8994
F	Lena	1	50	105	160	79	133	201	1.0000
	Baboon	1	56	182	89	115	158	135	1.0000
F'	Lena	1	105	79	160	133	201	50	1.0000
	Baboon	1	56	115	182	89	158	135	1.0000
Q	Lena	1	79	160	50	133	201	105	1.0000
	Baboon	1	158	115	89	182	135	56	1.0000

list the segmentation metrics ρ, F, F' and Q obtained in the segmentation of the test images of different dimensions and number of classes. The **boldfaced** values in Tables 3.13, 3.14, 3.15, 3.16, 3.17 and 3.18 signify the best metrics obtained with the FCM algorithm.

Table 3.16 FCM guided class boundaries and evaluated segmentation quality measures for evaluation functions ρ, F, F' and Q for 8 classes for 128×128 sized images

EF	Image	Set	c_1	c_2	c_3	c_4	c_5	c_6	c_7	c_8	η
ρ	Lena	1	73	152	201	95	115	133	173	51	**0.8478 (1)**
		2	201	51	132	95	173	152	114	72	0.8477 (2)
	Baboon	1	163	62	103	133	118	147	184	86	0.8734 (2)
		2	**86**	**103**	**62**	**134**	**147**	**163**	**184**	**118**	**0.8735 (1)**
	Synthetic	1	47	0	98	127	77	159	227	191	**0.9523 (1)**
		2	201	0	80	62	98	43	127	160	0.9430 (2)
	$MRI1$	1	48	72	25	118	105	129	91	0	**0.9196 (1)**
		2	72	48	118	129	91	0	25	105	0.9196 (1)
	$MRI2$	1	62	231	81	43	0	145	98	23	**0.8033 (1)**
		2	231	62	23	43	81	145	98	0	0.8033 (1)
F	Lena	1	97	120	51	202	138	174	154	74	**0.9814 (1)**
		2	152	173	201	51	132	114	72	95	1.0000 (2)
	Baboon	1	**184**	**163**	**134**	**103**	**62**	**86**	**119**	**148**	**0.9939 (1)**
		2	184	147	133	85	102	117	61	163	1.0000 (2)
	Synthetic	1	**159**	**227**	**0**	**77**	**127**	**191**	**98**	**47**	**0.9719 (1)**
		2	98	127	160	80	62	43	0	201	1.0000 (2)
	$MRI1$	1	**91**	**129**	**48**	**105**	**25**	**0**	**118**	**71**	**0.9938 (1)**
		2	91	24	0	70	117	129	47	104	1.0000 (2)
	$MRI2$	1	**0**	**81**	**98**	**62**	**231**	**42**	**23**	**145**	**0.9927 (1)**
		2	37	76	0	19	98	143	57	231	1.0000 (2)
F'	Lena	1	152	133	173	202	51	116	96	73	1.0000 (2)
		2	**138**	**202**	**175**	**154**	**97**	**120**	**51**	**74**	0.9855 (1)
	Baboon	1	166	89	64	137	108	184	124	152	1.0000 (2)
		2	**87**	**134**	**148**	**184**	**164**	**104**	**62**	**120**	0.9979 (1)
	Synthetic	1	**160**	**43**	**98**	**201**	**127**	**80**	**0**	**62**	**1.0000 (1)**
		2	201	160	127	98	62	0	43	80	1.0000 (1)
	$MRI1$	1	105	92	25	129	0	72	118	48	**1.0000 (1)**
		2	25	118	92	48	105	0	72	129	1.0000 (1)
	$MRI2$	1	0	231	38	97	143	59	78	21	1.0000 (2)
		2	**38**	**59**	**143**	**78**	**0**	**231**	**97**	**21**	0.9987 (1)
Q	Lena	1	**154**	**174**	**74**	**119**	**137**	**202**	**51**	**97**	**0.4326 (1)**
		2	136	153	74	97	119	202	174	51	1.0000 (2)
	Baboon	1	86	184	163	147	134	103	118	62	1.0000 (2)
		2	**62**	**103**	**86**	**147**	**134**	**119**	**184**	**163**	0.9784 (1)
	Synthetic	1	0	201	43	80	98	160	62	127	1.0000 (2)
		2	**197**	**100**	**132**	**64**	**81**	**48**	**33**	**0**	**0.2150 (1)**
	$MRI1$	1	**116**	**102**	**46**	**0**	**24**	**129**	**67**	**88**	**0.6968 (1)**
		2	25	72	105	48	92	129	0	118	1.0000 (2)
	$MRI2$	1	**21**	**77**	**38**	**97**	**143**	**0**	**59**	**231**	**0.4085 (1)**
		2	144	41	98	80	0	61	22	231	1.0000 (2)

Table 3.17 FCM guided class boundaries and evaluated segmentation quality measures for evaluation functions ρ, F, F' and Q for 8 classes for 256×256 sized images

EF	Image	Set	c_1	c_2	c_3	c_4	c_5	c_6	c_7	c_8	η
ρ	Lena	**1**	**70**	**130**	**153**	**184**	**104**	**39**	**6**	**226**	**0.9183 (1)**
		2	130	184	70	39	6	104	226	153	0.9182 (2)
	Baboon	**1**	**168**	**86**	**137**	**61**	**186**	**121**	**151**	**105**	**0.9056 (1)**
		2	103	60	120	167	84	186	150	118	0.9051 (2)
F	Lena	**1**	**103**	**6**	**38**	**70**	**153**	**226**	**184**	**129**	**0.9997 (1)**
		2	226	153	39	185	104	70	130	6	1.0000 (2)
	Baboon	**1**	**61**	**86**	**186**	**105**	**137**	**152**	**122**	**168**	**0.9984 (1)**
		2	167	61	104	121	186	136	151	85	1.0000 (2)
F'	Lena	1	127	151	226	183	6	38	69	101	1.0000 (2)
		2	**39**	**153**	**226**	**70**	**104**	**6**	**130**	**185**	**0.9964 (1)**
	Baboon	**1**	**85**	**121**	**186**	**104**	**167**	**136**	**61**	**151**	**0.9978 (1)**
		2	151	105	186	168	86	61	137	121	1.0000 (2)
Q	Lena	1	152	127	69	6	38	183	102	226	1.0000 (2)
		2	**130**	**184**	**39**	**153**	**70**	**104**	**226**	**6**	**0.8223 (1)**
	Baboon	**1**	**104**	**151**	**86**	**168**	**137**	**121**	**61**	**186**	**0.8429 (1)**
		2	186	60	120	84	103	136	150	167	1.0000 (2)

Table 3.18 FCM guided class boundaries and evaluated segmentation quality measures for evaluation functions ρ, F, F' and Q for 8 classes for 512×512 sized images

EF	Image	Set	c_1	c_2	c_3	c_4	c_5	c_6	c_7	c_8	η
ρ	Lena	1	99	123	157	49	206	178	75	140	0.9514
	Baboon	1	132	97	116	167	50	185	149	76	0.9076
F	Lena	1	99	206	75	140	178	49	157	123	1.0000
	Baboon	1	132	185	76	149	97	50	116	167	1.0000
F'	Lena	1	49	157	178	99	140	123	74	206	1.0000
	Baboon	1	185	76	131	167	97	116	149	50	1.0000
Q	Lena	1	156	99	74	177	49	140	206	122	1.0000
	Baboon	1	185	97	76	167	116	131	149	50	1.0000

3.7.2 Multilevel Image Segmentation Outputs

In this section, the segmented multilevel output images obtained for the different classes, with the optimised approach vis-a-vis those obtained with the heuristically chosen class boundaries and the standard FCM algorithm, are presented for the four quantitative measures used.

(a)   **(b)**   **(c)**   **(d)**

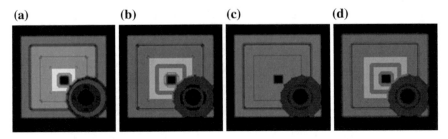

Fig. 3.9 8-class segmented 128 × 128 Synthetic image corresponding to the best EF value (from Table 3.4) obtained with OptiMUSIG activation function **a** ρ **b** F **c** F' **d** Q

(a)   **(b)**   **(c)**   **(d)**

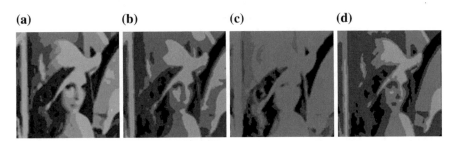

Fig. 3.10 8-class segmented 128 × 128 Lena image corresponding to the best EF value (from Table 3.4) obtained with OptiMUSIG activation function **a** ρ **b** F **c** F' **d** Q

(a)   **(b)**   **(c)**   **(d)**

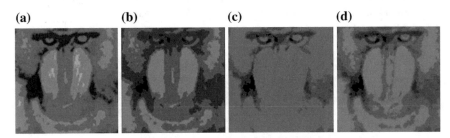

Fig. 3.11 8-class segmented 128 × 128 Baboon image corresponding to the best EF value (from Table 3.4) obtained with OptiMUSIG activation function **a** ρ **b** F **c** F' **d** Q

3.7.2.1 OptiMUSIG-guided segmented outputs

The segmented multilevel test images obtained with the MLSONN architecture using the OptiMUSIG activation function for the $K = 8$ class and different dimensions (128 × 128 and 256 × 256) corresponding to the best segmentation quality measures (ρ, F, F', Q) achieved, are shown in Figs. 3.9, 3.10, 3.11, 3.12, 3.13, 3.14 and 3.15.

The best outputs of the test images (512 × 512) obtained with the OptiMUSIG activation function with the evaluation function Q are shown in Figs. 3.16 and 3.17.

(a)   **(b)**   **(c)**   **(d)**

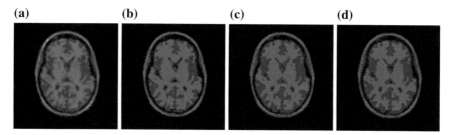

Fig. 3.12 8-class segmented 128 × 128 $MRI1$ image corresponding to the best EF value (from Table 3.4) obtained with OptiMUSIG activation function **a** ρ **b** F **c** F' **d** Q

(a)   **(b)**   **(c)**   **(d)**

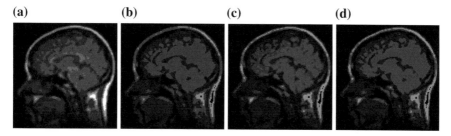

Fig. 3.13 8-class segmented 128 × 128 $MRI2$ image corresponding to the best EF value (from Table 3.4) obtained with OptiMUSIG activation function **a** ρ **b** F **c** F' **d** Q

3.7.2.2 MUSIG-guided segmented outputs

The segmented multilevel test images obtained with the MLSONN architecture characterised by the conventional MUSIG activation employing fixed class responses for $K = 8$ classes and different dimensions (128 × 128 and 256 × 256) yielding the best segmentation quality measures (ρ, F, F', Q) achieved, are shown in Figs. 3.18, 3.19, 3.20, 3.21, 3.22, 3.23 and 3.24.

The best outputs of the test images (512 × 512) obtained with the MUSIG activation function with the evaluation function Q are shown in Figs. 3.25 and 3.26

3.7.2.3 FCM guided segmented outputs

The segmented multilevel test images obtained with the standard FCM algorithm for $K = 8$ classes and different dimensions (128 × 128 and 256 × 256) yielding the best segmentation quality measures (ρ, F, F', Q) achieved, are shown in Figs. 3.27, 3.28, 3.29, 3.30, 3.31, 3.32 and 3.33.

The best outputs of the test images (512 × 512) obtained with the FCM activation function with the evaluation function Q are shown in Figs. 3.34 and 3.35.

A detailed scrutiny of different tables reveals that the presented OptiMUSIG activation function for the grey level test image segmentation outperforms the

(a)   **(b)**

(c)   **(d)**

Fig. 3.14 8-class segmented 256×256 Lena image corresponding to the best EF value (from Table 3.5) obtained with OptiMUSIG activation function **a** ρ **b** F **c** F' **d** Q

segmentation of the same test images by the conventional MUSIG activation function. A table-by-table comparison is narrated afterwards. The class boundaries and evaluated segmentation quality measures for different evaluation functions (ρ, F, F' and Q) for 6 classes for 128×128 sized images by the OptiMUSIG activation function, the MUSIG activation function and FCM algorithm are tabulated in Tables 3.1, 3.7 and 3.13, respectively. In these tables, results are reported for three test images. It is found that the segmentation quality measures by the OptiMUSIG activation function is awfully better than the other two algorithms. Now, we will throw some light on the Tables 3.2, 3.8 and 3.14. These tables depicted the class boundaries and evaluated segmentation quality measures for the same evaluation functions for 6 classes of 256×256 sized images. It is observed that the segmentation quality measures

(a)   **(b)**

(c)   **(d)**

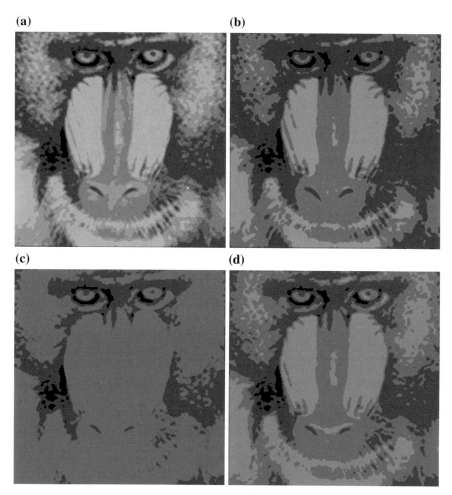

Fig. 3.15 8-class segmented 256×256 Baboon image corresponding to the best EF value (from Table 3.5) obtained with OptiMUSIG activation function **a** ρ **b** F **c** F' **d** Q

derived by the presented method is far better than other two methods. Tables 3.3, 3.9 and 3.15 are used to show the class boundaries and evaluated segmentation quality measures for the same evaluation functions for 6 classes of 512×512 sized images. The results reported in Tables 3.9 and 3.15 by the MUSIG activation function and FCM algorithm, respectively, are not as good as the same reported in Table 3.3. The same work has been done for the 8 classes and the results are reported in Tables 3.4, 3.5 and 3.6, 3.10, 3.11 and 3.12 and 3.16, 3.17 and 3.18 by the proposed method, the MUSIG activation function and the FCM algorithm, respectively. It is found that the segmentation quality measures derived by the proposed method is far better than other two methods when we make a detailed comparison of the results presented in Tables 3.4, 3.10 and 3.16. So, a conclusion can be made that the 8 class segmentation

Fig. 3.16 8-class segmented
512×512 Lena image
corresponding to the best Q
value (from Table 3.6)
obtained with OptiMUSIG
activation function

Fig. 3.17 8-class segmented
512×512 Baboon image
corresponding to the best Q
value (from Table 3.6)
obtained with OptiMUSIG
activation function

(a)   **(b)**   **(c)**   **(d)**

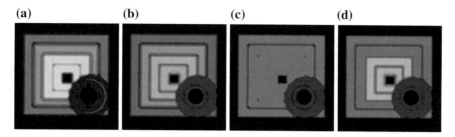

Fig. 3.18 8-class segmented 128×128 Synthetic image corresponding to the best EF value (from Table 3.10) obtained with MUSIG activation function **a** ρ **b** F **c** F' **d** Q

(a)   **(b)**   **(c)**   **(d)**

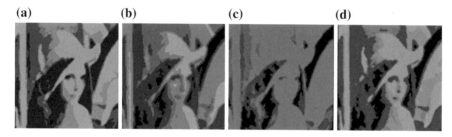

Fig. 3.19 8-class segmented 128×128 Lena image corresponding to the best EF value (from Table 3.10) obtained with MUSIG activation function **a** ρ **b** F **c** F' **d** Q

(a)   **(b)**   **(c)**   **(d)**

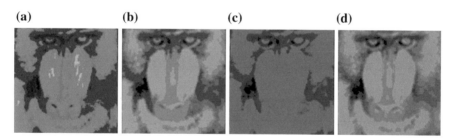

Fig. 3.20 8-class segmented 128×128 Baboon image corresponding to the best EF value (from Table 3.10) obtained with MUSIG activation function **a** ρ **b** F **c** F' **d** Q

(a)   **(b)**   **(c)**   **(d)**

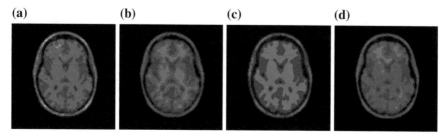

Fig. 3.21 8-class segmented 128×128 $MRI1$ image corresponding to the best EF value (from Table 3.10) obtained with MUSIG activation function **a** ρ **b** F **c** F' **d** Q

(a)   **(b)**   **(c)**   **(d)**

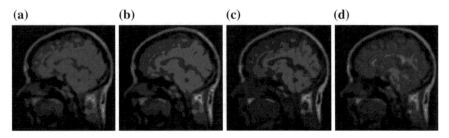

Fig. 3.22 8-class segmented 128×128 $MRI2$ image corresponding to the best EF value (from Table 3.10) obtained with MUSIG activation function **a** ρ **b** F **c** F' **d** Q

(a)   **(b)**

Fig. 3.23 8-class segmented 256×256 Lena image corresponding to the best EF value (from Table 3.11) obtained with MUSIG activation function **a** ρ **b** F **c** F' **d** Q

(a)   **(b)**

(c)   **(d)**

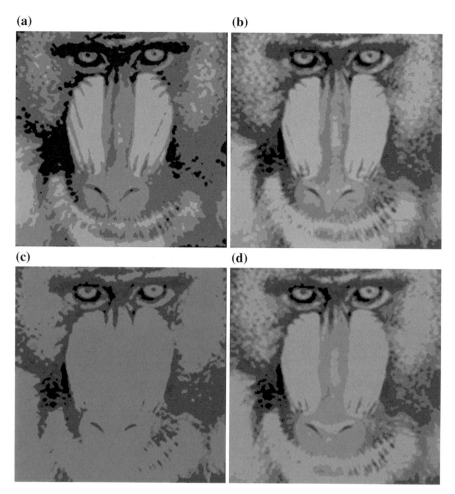

Fig. 3.24 8-class segmented 256 × 256 Baboon image corresponding to the best EF value (from Table 3.11) obtained with MUSIG activation function **a** ρ **b** F **c** F' **d** Q

done by the proposed method is better than the same done by the other two methods for 128 × 128 sized images. The segmentation done by the OptiMUSIG activation function is also better than the segmentation done by the conventional MUSIG activation function and the FCM algorithm for the 256 × 256 and 512 × 512 size images and the derived results are reported in the remaining tables.

From the results obtained, it is evident that the OptiMUSIG activation function outperforms its conventional MUSIG counterpart as well as the standard FCM algorithm as regards to the segmentation quality of the images for the different number of classes and dimensions of test images.

Fig. 3.25 8-class segmented 512×512 Lena image corresponding to the best Q value (from Table 3.12) obtained with MUSIG activation function .

Fig. 3.26 8-class segmented 512×512 Baboon image corresponding to the best Q value (from Table 3.12) obtained with MUSIG activation function

(a) **(b)** **(c)** **(d)**

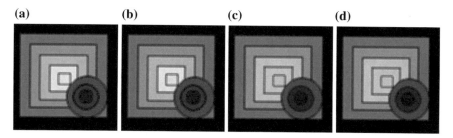

Fig. 3.27 8-class segmented 128 × 128 Synthetic image corresponding to the best EF value (from Table 3.16) obtained with FCM standard algorithm **a** ρ **b** F **c** F' **d** Q

(a) **(b)** **(c)** **(d)**

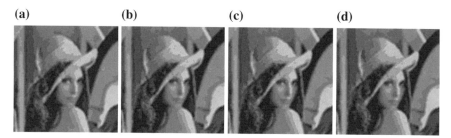

Fig. 3.28 8-class segmented 128 × 128 Lena image corresponding to the best EF value (from Table 3.16) obtained with FCM standard algorithm **a** ρ **b** F **c** F' **d** Q

(a) **(b)** **(c)** **(d)**

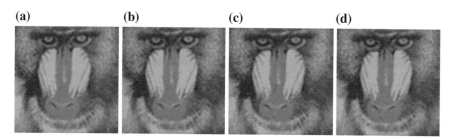

Fig. 3.29 8-class segmented 128 × 128 Baboon image corresponding to the best EF value (from Table 3.16) obtained with FCM standard algorithm **a** ρ **b** F **c** F' **d** Q

(a) **(b)** **(c)** **(d)**

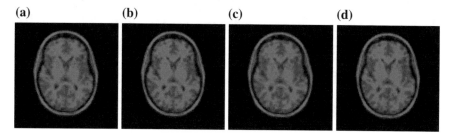

Fig. 3.30 8-class segmented 128 × 128 $MRI1$ image corresponding to the best EF value (from Table 3.16) obtained with FCM standard algorithm **a** ρ **b** F **c** F' **d** Q

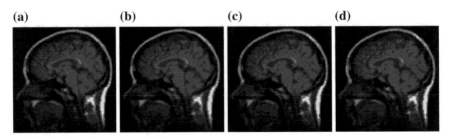

Fig. 3.31 8-class segmented 128×128 $MRI2$ image corresponding to the best EF value (from Table 3.16) obtained with FCM standard algorithm **a** ρ **b** F **c** F' **d** Q

Fig. 3.32 8-class segmented 256×256 Lena image corresponding to the best EF value (from Table 3.17) obtained with FCM standard algorithm **a** ρ **b** F **c** F' **d** Q

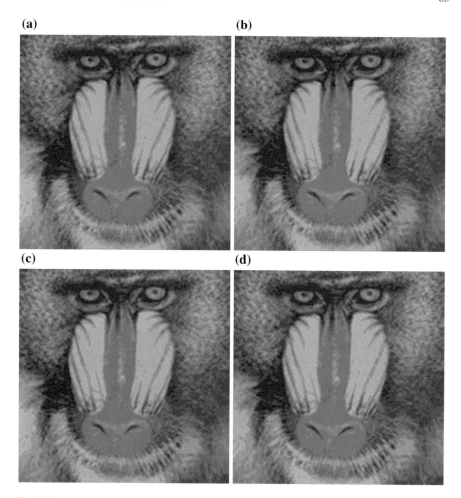

Fig. 3.33 8-class segmented 256×256 Baboon image corresponding to the best EF value (from Table 3.17) obtained with FCM standard algorithm **a** ρ **b** F **c** F' **d** Q

3.8 Discussions and Conclusion

A multilayer self-organising neural network (MLSONN) architecture is discussed is this chapter. The basis of induction of multiscaling capabilities in the network by resorting to a multilevel MUSIG activation function is also reported in this chapter.

The inherent limitation of the MUSIG activation function as regards to its reliance on fixed and heuristic class responses has been addressed in this chapter. An optimised version of the MUSIG activation function (the OptiMUSIG activation function) characterised by optimised class boundaries of the multilevel images, is proposed. A genetic algorithm-based optimization procedure with different objective functions (as measures of image segmentation quality) is used to derive the optimised class

Fig. 3.34 8-class segmented
512 × 512 Lena image
corresponding to the best Q
value (from Table 3.18)
obtained with FCM standard
algorithm

Fig. 3.35 8-class segmented
512 × 512 Baboon image
corresponding to the best Q
value (from Table 3.18)
obtained with FCM standard
algorithm

boundaries. Applications of the proposed OptiMUSIG activation function for the segmentation of real-life multilevel images show superior performance as compared to the MUSIG activation function with heuristic class boundaries. Furthermore, comparative analysis of the results obtained with the standard FCM algorithm also signifies the merit of the proposed approach.

Methods will be investigated to apply the proposed OptiMUSIG activation function for the segmentation of colour images as well. The next chapter of this book is intended in this direction.

Chapter 4
Self-supervised Colour Image Segmentation Using Parallel OptiMUSIG (ParaOptiMUSIG) Activation Function

4.1 Introduction

Colour image segmentation and analysis form a challenging proposition in the image processing arena owing to the nature and variety of data to be processed [258, 259]. Huge amount of complex computational efforts is required to process colour images as the underlying data exhibits information in primary colour components and their admixtures. Another challenging proposition faced in the processing of colour images is the nonlinearity in the representation of colours in the colour spectrum. Colour image processing and segmentation are classical examples of multichannel information processing [150, 197, 258]. Basically, a colour image is characterised either by information in three primary colour components, viz., red, green, and blue or in all their possible combinations. The primary colours as well as the combined colour information appear either with the minimum intensity value of 0 or with the maximum intensity value of 255. The colour spectrum of a true colour image is formed with all the possible combinations of intensity values from 0 to 255 for each of the primary colour components and their admixtures. Thus, processing and understanding of a colour image amount to processing of the combinations of the full strengths of the primary colour component information in a pure colour image and processing of all the combinations of these colour components in a true colour image. Colour image segmentation finds wide use in the field of remote sensing [258, 260], multimedia-based information systems [258, 261], GIS [258, 262], and multispectral data management systems [258, 263], to name a few.

A good survey of different type of classical approaches of image segmentation and analysis have been accounted in the literature [93, 258, 264]. Classical image segmentation approaches can be generally relegated into three major categories, i.e., feature space based segmentation, image domain based segmentation, and graph-based segmentation. In feature space based techniques, image segmentation is achieved by capturing the global characteristics of the image through the selection and calculation of the image features, viz. grey level, colour level, texture, to name a few [93, 265].

© Springer International Publishing AG 2016
S. De et al., *Hybrid Soft Computing for Multilevel Image and Data Segmentation*, Computational Intelligence Methods and Applications, DOI 10.1007/978-3-319-47524-0_4

Histogram thresholding technique is a very common technique in this arena. The histogram thresholding of a colour image relies on the peaks or valleys in three colour histograms or a three-dimensional histogram. A segmentation method to extract regions within a image has been presented by thresholding the HSV histograms of a colour image [258, 266]. Split and merge, region-growing, edge detection, etc., are the examples of image domain based segmentation techniques based on the strategy of spatial grouping [93, 258]. The working principle of region-based technique is based on the region entities or on the spatial information of the regions. A region-growing image segmentation approach, named JSEG, has been proposed by Deng et al. [267]. This method comprises colour quantization on the basis of the pixel labels of the colour images and spatial segmentation. The image segmentation criterion of this method uses the colour map employed on the J-image and the resultant high and low values in J-images denotes the possible boundaries and centres of the regions. Ruzon and Tomasi [268] designed a compass operator to compare colour distributions and detect edges in colour images. In this method, the colour distribution of each pixel is compared with both side of the region using the Earth Mover's distance (EMD) [269], a robust histogram-matching method. But the computational running time of this method is high [270]. Normalised cut [143] is a good approach of the third category, viz. graph-based image segmentation. Several image segmentation approaches using the normalised cut are already reported in the Sect. 2.2 of Chap. 2. The detailing of an image is overlooked by the normalised cut. A hybrid segmentation algorithm which mixes the prior shape information with normalised cut is presented in [271]. In this process, the normalised cut can segment the target object from a noisy image effectively with the help of shape information [271, 272]. The image segmentation techniques using classical approaches always demand some a priori knowledge regarding the feature space and its distribution in the image data. These are the limitations of these approaches. But the nonclassical approaches viz. the neuro-fuzzy-genetic approaches work on the underlying data regardless of the distribution and operating parameters.

In the previous chapter, it has been pointed out that the conventional MUSIG activation function to segment multilevel greyscale images in connection with the conventional multilayer self-organising neural network (MLSONN) architecture is constrained to fetch the image information to generate the function. The class levels to generate that activation function are predefined and they do not vary from image to image. Moreover, this activation function uses uniform thresholding parameters. The main reason behind this limitation is that it intends that the underlying information of an image are homogeneous in nature though the real situation is totally opposite.

The main focus of the previous chapter was to bring about a functional change in the operational characteristics of the MUSIG activation function, thereby inducing inherent image information to generate the activation function. The approach undertaken for this purpose was targeted at achieving the optimised class levels from the underneath image content. The much sought functional modification of the MUSIG activation function was effected by incorporating the optimised class levels and the modified activation function is denoted as the optimised MUSIG (OptiMUSIG) activation function [128, 134, 136, 172, 177]. The OptiMUSIG activation function [128,

134, 136, 172, 177] has been generated by a genetic algorithm based optimization procedure. This function integrates the heterogeneous image information content in the MUSIG activation function to segment multilevel greyscale images.

In this chapter, we have extended the functionality of the OptiMUSIG activation function to the colour domain so that it is able to segment colour images. The natural extensions of binary and multilevel images are the colour images. A pure colour image presents information in three primary colour components and their admixtures. A pure colour image has these **R**ed, **G**reen and **B**lue colour components and their admixtures with two intensity levels viz. 0 and 255. Therefore, a pure colour image can be considered as an extension of a binary image with respect to the intensity levels exhibited by the constituent colour components and admixtures. In true colour images, intensity information is evidenced in all the possible 256 levels (0–255) of these primary colour components and their admixtures. Thus, the true colour image can be intended as an extension of a multilevel image which also displays intensity levels from 0 (perfectly black) to 255 (perfectly white). Keeping these extensions in mind, colour image processing can be accomplished independently by simply replicating the binary/multilevel image processing paradigm for the three primary colour components. Processing of these independent colour components can be attempted in parallel. This idea of parallel and independent processing of colour component image information can be applied to the parallel version of the MLSONN architecture, noted as the parallel self-organising neural network (PSONN) architecture [195, 196]. The parallel architecture consists of an input (source) layer of image information, three single MLSONN architectures, individual one committed to the segmentation of the different primary colour component information in a colour image and one final output (sink) layer for fusion of the segmented colour component images. This PSONN network architecture is able to segment the colour images by way of component level processing of incident image information. This network architecture segments the primary colour component information in parallel and simultaneously. In this respect, a parallel version of the OptiMUSIG (ParaOptiMUSIG) activation function [193, 258, 273] is employed in the PSONN network architecture to segment the colour images.

This chapter is aimed to segment true colour images into different number of classes with the help of optimised class boundaries for individual colour components. The proposed genetic algorithm based optimization techniques has been applied to generate these colour components in parallel. The individual colour component is transferred to individual SONNs in the PSONN architecture [195, 196] by differentiating the input true colour images into different colour components. However, different activation functions are applied for different SONNs. The parallel version of the optimised MUSIG (ParaOptiMUSIG) activation function [193, 258, 273] for the PSONN are designed with the optimised class boundaries for all colour components. These optimised class boundaries are determined from the image context. Finally, the sink layer of the PSONN architecture fuses the processed colour component information into a processed colour image. The number of target classes in the resultant segmented image corresponds to the number of levels in the multilevel activation function.

The application of the proposed genetic algorithm based ParaOptiMUSIG activation function approach is demonstrated using true colour version of the Lena and Baboon images. A colour image of a human brain using a magnetic resonance imaging (MRI) machine [36] is also applied to establish the utility of the proposed approach. The image segmentation evaluation index like the standard measure of correlation coefficient (ρ) [72], entropy-based index [170, 178], or quantitative-based index [170, 256] are very much efficient to assess the quality of the image segmentation results. A higher value of the correlation coefficient and lower quantitative value or entropy value signify better segmentations. The standard measure of correlation coefficient (ρ) [72] and three measures, viz., F due to Liu and Yang [255] and F' and Q due to Borsotti et al. [256] have been applied to evaluate the segmentation efficiency of the proposed approach. Results of segmentation using the proposed ParaOptiMUSIG activation function show better performance over the conventional MUSIG activation function employing heuristic class responses.

The organisation of this chapter is as follows: a brief description of the PSONN architecture and its operation is given in Sect. 4.2. Section 4.3 describes the proposed Parallel optimised MUSIG (ParaOptiMUSIG) activation function. The algorithm for the designing of the ParaOptiMUSIG activation function is also presented in this section. In Sect. 4.4, a description of the proposed methodology for true colour image segmentation is presented. A comparative study of the results of segmentation of the test images by the proposed approach vis-a-vis by the conventional MUSIG activation function is illustrated in Sect. 4.5. Section 4.6 concludes the chapter with future directions of research.

4.2 Parallel Self-Organising Neural Network (PSONN) Architecture

The multilayer self-organising neural network (MLSONN) [3] is a feedforward neural network architecture that operates in a self-supervised manner. It consists of an input layer, a hidden layer, and an output layer of neurons. A detailed study of the architecture and operation of the MLSONN architecture is demonstrated in Chap. 3. The three-layer self-organising neural network (TLSONN) architecture is a smaller version of the generalised MLSONN architecture. It comprises a single hidden layer of neurons apart from the input and output layers.

Three independent single three-layer self-organising neural network (TLSONN) architecture for component level processing constitute the parallel version of the self-organising neural network (PSONN) [195, 196] in addition with a source layer for inputs to the network and a sink layer for generating the final network output. This network can be employed for extracting pure colour objects from a pure colour image [196]. The source layer disperses the primary colour component information of true colour images into the three parallel SONN architectures. After processing of each colour component at these three parallel SONN architectures, the sink layer fuses and generates the final pure colour output images. However, the three parallel self-organising neural network architectures operate in a self-supervised mode on multiple shades of colour component information [196]. The linear indices of fuzziness of the colour component information obtained at the respective output layers are employed to calculate the system errors. The respective interlayer interconnection weights using the standard backpropagation algorithm [37, 47] are adjusted with these system errors. This method of self-supervision proceeds on until the system errors at the output layers of the three independent SONNs fall below some tolerable limits. The extracted colour component outputs are produced by the corresponding output layers of three independent SONN architectures. Finally, the extracted pure colour output image is developed by mixing the segmented component outputs at the sink layer of the PSONN network architecture [195, 196].

Basically, the object and the background regions in the input image are segregated on the basis of the intensity features of the pixels of different regions. The PSONN [195, 196] network architecture is not capable to segment multilevel input images, i.e., inputs which consist of different heterogeneous shades of image pixel intensity levels. This is solely due to the fact that the constituent primitives/SONNs resort to the use of the standard bilevel sigmoidal activation functions. However, the problem of these network architectures can be facilitated by resorting to a modified PSONN network architecture. For this reason, Bhattacharyya et al. [72] introduced a multilevel version of the generalised bilevel sigmoidal activation function. The novelty of the multilevel sigmoidal (MUSIG) activation [72] function consists of its ability to map inputs into multiple output levels and capable to segment true colour images efficiently. The detailed review of the architecture and operational characteristics of the PSONN network architecture can be found in the articles [195, 196].

4.3 Parallel optimised Multilevel Sigmoidal (ParaOptiMUSIG) Activation Function

The main objective of this chapter is to segment true colour images, using a PSONN architecture [195, 196] by optimised embedding functional characteristics in the constituent SONN primitives obtained from the true colour image context. The inherent colour components in the colour image is decomposed into R, G, and B colour components to process those colour components by the three coupled SONNs. Each colour component of a particular pixel is segmented into different class levels in parallel. After the segmentation, the individual segmented class levels of the individual colour components for that particular pixel are combined to get the segmented image.

This problem can be extended for multidimensional datasets as well, where the segmentation of every feature can be carried out similarly in parallel. The resultant segmented value of that datapoint is thereby obtained through the combination of all the features in the parallel segmentation process. It is worth mentioning at this point that the different class levels for the segmentation of such a dataset can be usually determined heuristically from the histograms of the k features of the dataset, ignoring the heterogeneity of the underlying data information. But, since the datapoints in a real-life dataset generally exhibit a fair amount of heterogeneity and the class levels generally differ from one dataset to another, such a heuristic approach does not always reflect the true essence of information content underneath. Thus, the variation of datapoints of that dataset can be incorporated in the segmentation by accommodating the heterogeneity in the segmentation process in one way or the other.

Over the years, scientists have resorted to several techniques for segmenting datasets in various ways. Neural networks, discussed in the previous section, have been found to be quite efficient in this regard as well when fed with the required class levels of the segmentation process. In particular for a kth dimensional dataset, one can either employ an ensemble of competitive neural networks [274, 275] with a voting mechanism to decide on the final segmentation procedure or a collection of neural networks connected in parallel [195] assigned for each and every feature of the datapoints in the dataset. In either case, the activation/transfer functions of the neural networks play important roles for the performance of the neural network architectures. The standard bilevel sigmoidal activation function is one of the commonly used activation function in neural networks but this function is unable to classify the data features into multiple levels as the function generates the bipolar responses [0(low)/1(high)] corresponding to input information. The multilevel sigmoidal (MUSIG) activation function [72] is able to produce multilevel scales of grey to incident input multilevel information and efficient to segment the multilevel greyscale images. The main drawback of this activation function is that the class responses (c_γ) are selected randomly from the feature histogram of the datapoints and this function is totally independent of the nature and distribution of the operated dataset. The class levels of the datapoints can be incorporated in the characteristics neuronal activations by the optimised class boundaries derived from the dataset. The optimised class boundaries can be generated by the genetic algorithm as this algorithm is stochastic in nature and the operators of this algorithm have inherent characteristics that generate random class responses (c_γ) from the operated dataset. The optimised class responses (c_γ) are treated as the optimised class boundaries for the operated dataset and are applied in the optimised version of the MUSIG activation (OptiMUSIG) [172] function. The detailed study of the MUSIG activation function and the OptiMUSIG activation function is presented in Sect. 3.4 of Chap. 3.

This OptiMUSIG activation function can be used in each constituent SONN primitive of PSONN [195, 196] with appropriate optimised parameter settings. The parallel representation of the OptiMUSIG (ParaOptiMUSIG) activation function is denoted as [193, 258, 273]

$$f_{\text{ParaOptiMUSIG}} = \sum_{t \in \{t_1, t_2, \dots, t_n\}} f_{t_{\text{OptiMUSIG}}} \qquad (4.1)$$

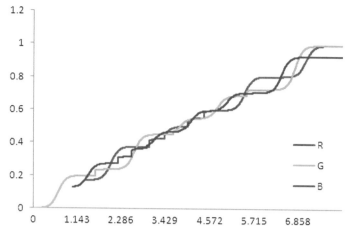

Fig. 4.1 ParaOptiMUSIG activation function applied for R, G, B colour components with $K = 8$ classes

where $\{t_1, t_2, ..., t_n\}$ denotes the different layers of the parallel self-organising neural network (PSONN) [196, 195] architecture and $f_{t_{\text{OptiMUSIG}}}$ denotes the Opti-MUSIG activation function for one layer of the network [193, 258, 273]. Here, the $\sum$ sign denotes the collection of the OptiMUSIG functions of different layers. The optimised class boundaries for different OptiMUSIG activation functions of different layers are generated by the genetic algorithm in parallel. A designed ParaOptiMUSIG activation function for the different R, G, B colour components for $K = 8$ classes with the optimised colour levels to process the individual colour components is shown in Fig. 4.1. Similarly, Fig. 4.2 shows three different MUSIG activation functions applied in parallel to process the different R, G, B colour components for $K = 8$ classes of the constituent component networks. From Figs. 4.1 and 4.2, it is evident that the transition lobes of the constituent R, G, B components of the ParaOptiMUSIG activation function in Fig. 4.1 are sharply and distinctly spread from each other as compared to the heuristically designed R, G, B collections in the MUSIG activation function in Fig. 4.2. These imply a broader operating spectrum of the ParaOptiMUSIG activation function essentially arising out of the incorporation of the image information content in the optimization and thresholding process.

4.4 ParaOptiMUSIG Activation Function Based Colour Image Segmentation Scheme

This section describes the proposed true colour image segmentation technique by ParaOptiMUSIG activation function with a PSONN architecture in detail. The different phases of this technique are discussed elaborately in the following subsections and the flow diagram [193, 258] is depicted in Fig. 4.3.

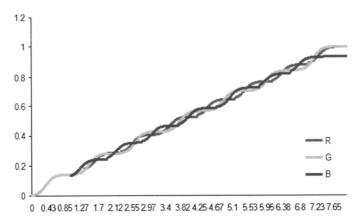

Fig. 4.2 Three MUSIG activation function applied in parallel for R, G, B colour components for $K = 8$ classes

4.4.1 Optimised Class Boundaries Generation for True Colour Images

In this most important phase of the proposed approach, a GA-based optimization procedure is applied to generate the optimised class boundaries $(c_{\gamma_{opt}})$ of the proposed ParaOptiMUSIG activation function. The procedure employed in this phase is as follows.

4.4.1.1 Input Phase

The pixel intensity levels of the true colour image and the number of classes (K) to be segmented are supplied as inputs to this GA based optimization procedure.

4.4.1.2 Chromosome Representation and Population Generation

A binary encoding technique for the chromosomes is applied for representing the optimised class boundaries from the input true colour image information content. Each pixel intensity of the true colour image information is differentiated into three colour components, viz. red, green, and blue colour components. Three different chromosome pools are produced for the three individual colour components. Each chromosome pool is used to generate the optimised class levels for the individual colour component. Randomly selected binary combinations of eight bits represent the class boundary level of the segmentation. If the image is segmented into K segments in the image, the size of the chromosome equals $K \times 8 \times 3$ bits and the class boundaries for the individual colour component encoded in a chromosome in the

Fig. 4.3 Flow diagram of the proposed approach

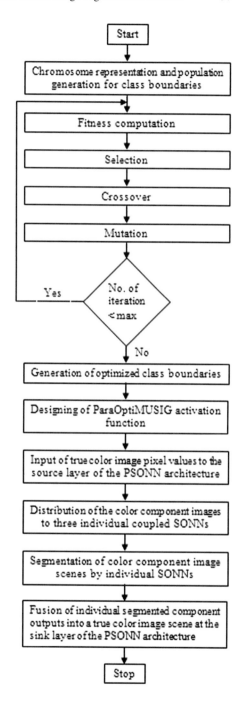

initial population are randomly chosen to obtain K distinct class boundaries from the image content. A population size of 200 has been employed for this treatment.

4.4.1.3   Fitness Computation

In this phase, four segmentation evaluation criteria (ρ, F, F', Q) given in Eqs. 3.25, 3.28, 3.29 and 3.30 respectively, are applied as the fitness functions. These functions are applied to evaluate the quality of the segmented images in this genetic algorithm based optimization procedure.

4.4.1.4   Genetic Operators

In the selection phase, a proportionate fitness selection operator is employed to select the reproducing chromosomes. The selection probability of the ith chromosome is evaluated using the Eq. 3.31. Subsequently, the crossover and mutation operators are applied to evolve a new population. In this approach, the crossover probability is equal to 0.8. A single point crossover operation is applied to generate the new pool of chromosomes. The mutation probability is taken as 0.1 in this approach. After mutation, the child chromosomes are propagated to form the new generation.

4.4.2   ParaOptiMUSIG Activation Function Design

The ParaOptiMUSIG activation function is designed by the optimised class boundaries ($c_{\gamma_{opt}}$) those are obtained from the previous phase. The optimised class boundaries for individual colour component in the selected chromosomes are applied to generate the individual OptiMUSIG activation function for that colour component, viz. the class boundaries for the red component is employed to generate OptiMUSIG activation function for red and so on. The ParaOptiMUSIG function is derived by collection as the individual OptiMUSIG functions generated for the individual colour components using Eq. 4.1.

4.4.3   Input of True Colour Image Pixel Values to the Source Layer of the PSONN Architecture

In this phase, the source layer of the PSONN architecture is fed with the pixel intensity levels of the true colour image. The number of neurons in the source layer of the PSONN architecture is same as the number of pixels in the processed image.

4.4.4 Distribution of the Colour Component Images to Three Individual SONNs

The individual primary colour components are differentiated from the pixel intensity levels of the input true colour image and the three individual three-layer component SONNs are used for these individual primary colour components, viz. the red component is applied to one SONN, the green component to another SONN and the remaining SONN accepts the blue component information at their respective input layers. The fixed interconnections of the respective SONNs with the source layer are responsible for this scenario.

4.4.5 Segmentation of Colour Component Images by Individual SONNs

The corresponding SONN architecture channelised by the projectedParaOptiMUSIG activation function at the individual primitives/neurons is applied to segment the individual colour components of the true colour images. Depending on the number of transition lobes of the ParaOptiMUSIG activation function the neurons of the different layers of individual three-layer SONN architecture yield different input colour component level responses. The system errors are evaluated by the subnormal linear index of fuzziness (v_{l_s}) [72] at the corresponding output layers of the individual SONNs since the network has no a priori knowledge about the outputs. The standard backpropagation algorithm [37, 47] is then applied to adjust the interconnection weights between the different layers to minimise the errors. The respective output layers of the independent SONNs render the final colour component images when the self-supervision of the corresponding networks achieve stabilisation.

4.4.6 Fusion of Individual Segmented Component Outputs into a True Colour Image at the Sink Layer of the PSONN Architecture

The segmented outputs derived at the three output layers of the three independent three-layer SONN architectures are fused at the sink layer of the PSONN architecture to deduce the segmented true colour image. The number of segments is a combination of the number of transition lobes of the designed ParaOptiMUSIG activation functions applied during component level segmentation.

Basically, *RGB* colour model in applied in this process. Any colour expressed in an *RGB* space is an admixture of the three primary colour components, viz., red, green, and blue. As we know, *RGB* colour model is an additive colour model, i.e., any other colour is generated by the weighted sum of the primary colours. So, a colour C represented in RGB components as

$$C = R\mathbf{R} + G\mathbf{G} + B\mathbf{B} \qquad (4.2)$$

where, the values of R, G and B are ranged within 0 to 1. The RGB colour space can be represented by a unit cube placed at the origin in its standard position. In this work, a standard RGB colour function is applied to generate the colour pixel from the R, G and B colour components.

4.5   Experimental Results

In this chapter, two real-life true colour images viz. Lena and Baboon each of dimension 256×256 and a colour image of a human brain using a magnetic resonance imaging (MRI) machine [36] of dimension 170×170 have been applied to demonstrate the purported true colour image segmentation approach using the ParaOptiMUSIG activation function in connection with the PSONN architecture. The ParaOptiMUSIG activation function has been prepared with a fixed slope, $\lambda = \{2, 4\}$ for $K = \{4, 6, 8\}$ classes. In this chapter, the results are reported with a fixed slope $\lambda = 4$ in combination with $\{6, 8\}$ classes for Lena and Baboon images and with 8 classes for colour MRI image. In the following Sects. 4.5.1.1 and 4.5.2.1, the quantifiable performance analysis of the proposed GA-based ParaOptiMUSIG activation function and the corresponding segmented outputs are depicted, respectively. The segmentation derived by the conventional MUSIG activation function with same number of class responses and with heuristic class levels is compared with the proposed approach. Segmentation efficiency of the conventional MUSIG activation function as regards to its efficacy in the segmentation of true colour test images are elaborated in the Sects. 4.5.1.2 and 4.5.2.2.

4.5.1   Quantitative Performance Analysis of Segmentation

Four evaluation functions (ρ, F, F' and Q) have been employed to demonstrate the quantitative measures of the effectivity of the proposed ParaOptiMUSIG and the conventional MUSIG activation functions for $K = \{6, 8\}$ in this section. The ParaOptiMUSIG activation function based experimental evaluation results have been discussed in Sect. 4.5.1.1 A and the corresponding results deduced with the conventional fixed class response based MUSIG activation function are illustrated in Sect. 4.5.1.2 B.

4.5.1.1   ParaOptiMUSIG Activation Function Based Segmentation Evaluation

The genetic algorithm based optimization procedure generates the optimised sets of class boundaries $c_{\gamma_{opt}}$ on the basis of four evaluation function (ρ, F, F' and Q) for different number of classes. These are tabulated in Tables 4.1, 4.2, 4.3, 4.4, 4.5, 4.6, 4.7, 4.8, 4.9, 4.10, 4.11, 4.12, 4.13, 4.14, 4.15, 4.16, 4.17, 4.18, 4.19 and 4.20 for the three test images. The evaluation functions (EF_{op}) which are treated as the fitness functions in this proposed approach are shown in the first columns of these tables and in the third column of the tables, the optimised set of class boundaries are accounted. Two sets of results per evaluation function of each test image are reported. The last columns of the tables show the quality measures η [graded on a scale of **1** (best) to **2** (worst)] obtained by the segmentation of the test images based on the corresponding set of optimised class boundaries. It is to be noted that the evaluation function values are reported in normalised form in this chapter. The **boldfaced** result in each table denotes the best values obtained by the proposed approach for easy reckoning. These quality measures are applied to compare with the quality measures those are derived with the heuristically selected class boundary based conventional MUSIG activation function.

4.5.1.2   Segmentation Evaluation by MUSIG Activation Function

In the same manner, the first column of the Tables 4.1, 4.2, 4.3, 4.4, 4.5, 4.6, 4.7, 4.8, 4.9, 4.10, 4.11, 4.12, 4.13, 4.14, 4.15, 4.16, 4.17, 4.18, 4.19 and 4.20 are tabulated with the evaluation functions (EF_{fx}) corresponding to the heuristically selected class boundaries with the conventional MUSIG activation function. The corresponding

Table 4.1 Optimised and fixed class boundaries and evaluated segmentation quality measures, ρ for 6 classes of Lena image

EF	Set	Colour levels	η
ρ_{op}	1	$R = \{43, 99, 146, 171, 191, 255\}$	**0.8994 (1)**
		$G = \{0, 33, 124, 12, 223, 255\}$	
		$B = \{32, 66, 91, 129, 167, 238\}$	
	2	$R = \{43, 91, 100, 113, 195, 255\}$	0.8960 (2)
		$G = \{0, 45, 87, 98, 135, 255\}$	
		$B = \{32, 69, 94, 121, 165, 238\}$	
ρ_{fx}	1	$R = \{43, 45, 85, 135, 200, 255\}$	**0.8622 (1)**
		$G = \{0, 44, 87, 137, 210, 255\}$	
		$B = \{32, 41, 82, 134, 212, 238\}$	
	2	$R = \{43, 45, 90, 125, 178, 255\}$	0.8614 (2)
		$G = \{0, 56, 110, 140, 186, 255\}$	
		$B = \{32, 55, 105, 125, 195, 238\}$	

Table 4.2 Optimised and fixed class boundaries and evaluated segmentation quality measures, F for 6 classes of Lena image

EF	Set	Colour levels	η
F_{op}	1	$R = \{$**43, 78, 114, 239, 251, 255**$\}$	**0.522 (1)**
		$G = \{$**0, 18, 31, 163, 203, 255**$\}$	
		$B = \{$**32, 48, 162, 183, 193, 238**$\}$	
	2	$R = \{43, 71, 77, 103, 212, 255\}$	0.527 (2)
		$G = \{0, 121, 150, 195, 210, 255\}$	
		$B = \{32, 53, 157, 176, 204, 238\}$	
F_{fx}	1	$R = \{43, 90, 100, 115, 140, 255\}$	1.000 (2)
		$G = \{0, 20, 130, 150, 160, 255\}$	
		$B = \{32, 70, 105, 120, 160, 238\}$	
	2	$R = \{$**43, 125, 130, 210, 230, 255**$\}$	**0.707 (1)**
		$G = \{$**0, 80, 100, 135, 200, 255**$\}$	
		$B = \{$**32, 140, 155, 200, 230, 238**$\}$	

Table 4.3 Optimised and fixed class boundaries and evaluated segmentation quality measures, F' for 6 classes of Lena image

EF	Set	Colour levels	η
F'_{op}	1	$R = \{43, 72, 172, 206, 223, 255\}$	0.571 (2)
		$G = \{0, 60, 174, 196, 223, 255\}$	
		$B = \{32, 66, 82, 157, 163, 238\}$	
	2	$R = \{$**43, 78, 95, 132, 233, 255**$\}$	**0.501 (1)**
		$G = \{$**0, 21, 43, 69, 132, 255**$\}$	
		$B = \{$**32, 48, 54, 118, 134, 238**$\}$	
F'_{fx}	1	$R = \{$**43, 80, 130, 180, 205, 255**$\}$	**0.734 (1)**
		$G = \{$**0, 25, 35, 135, 159, 255**$\}$	
		$B = \{$**32, 85, 135, 175, 179, 238**$\}$	
	2	$R = \{43, 75, 120, 140, 182, 255\}$	1.000 (2)
		$G = \{0, 50, 85, 90, 177, 255\}$	
		$B = \{32, 60, 150, 180, 190, 238\}$	

quality measures assessed after the segmentation process along with the **boldfaced** best results are also shown alongside. It is quite observable from these tables that the fitness values derived by the ParaOptiMUSIG activation function are better than those obtained by the conventional MUSIG activation function.

Table 4.4 Optimised and fixed class boundaries and evaluated segmentation quality measures, Q for 6 classes of Lena image

EF	Set	Colour levels	η
Q_{op}	1	$R = \{43, 130, 171, 187, 204, 255\}$	**0.608 (1)**
		$G = \{0, 97, 111, 129, 172, 255\}$	
		$B = \{32, 81, 99, 105, 197, 238\}$	
	2	$R = \{43, 147, 188, 204, 221, 255\}$	0.628 (2)
		$G = \{0, 117, 131, 149, 193, 255\}$	
		$B = \{32, 50, 96, 119, 124, 238\}$	
Q_{fx}	1	$R = \{43, 45, 85, 135, 200, 255\}$	1.000 (2)
		$G = \{0, 44, 87, 137, 210, 255\}$	
		$B = \{32, 41, 82, 134, 212, 238\}$	
	2	$R = \{43, 80, 17, 206, 220, 255\}$	**0.828 (1)**
		$G = \{0, 115, 176, 192, 207, 255\}$	
		$B = \{32, 85, 130, 150, 180, 238\}$	

Table 4.5 Optimised and fixed class boundaries and evaluated segmentation quality measures, ρ for 6 classes of Baboon image

EF	Set	Colour levels	η
ρ_{op}	1	$R = \{0, 29, 92, 97, 146, 255\}$	0.9531 (2)
		$G = \{0, 101, 154, 168, 192, 255\}$	
		$B = \{0, 61, 88, 128, 165, 255\}$	
	2	$R = \{0, 133, 198, 215, 250, 255\}$	**0.9562 (1)**
		$G = \{0, 64, 155, 186, 192, 255\}$	
		$B = \{0, 61, 100, 130, 176, 255\}$	
ρ_{fx}	1	$R = \{0, 42, 72, 206, 223, 255\}$	**0.9296 (1)**
		$G = \{0, 40, 174, 196, 223, 255\}$	
		$B = \{0, 36, 82, 157, 163, 255\}$	
	2	$R = \{0, 78, 90, 192, 240, 255\}$	0.8817 (2)
		$G = \{0, 30, 95, 185, 200, 255\}$	
		$B = \{0, 10, 102, 210, 233, 255\}$	

4.5.2   *True Colour Image Segmentation Outputs*

The segmented true colour output images obtained with the proposed GA-based optimised approach and those obtained with the randomly chosen class boundaries according to the quantitative measure are demonstrated in this section.

Table 4.6 Optimised and fixed class boundaries and evaluated segmentation quality measures, F for 6 classes of Baboon image

EF	Set	Colour levels	η
F_{op}	1	$R = \{0, 20, 39, 61, 204, 255\}$	**0.491 (1)**
		$G = \{0, 15, 53, 187, 200, 255\}$	
		$B = \{0, 10, 169, 206, 213, 255\}$	
	2	$R = \{0, 16, 35, 81, 238, 255\}$	0.495 (2)
		$G = \{0, 29, 41, 159, 194, 255\}$	
		$B = \{0, 10, 186, 218, 233, 255\}$	
F_{fx}	1	$R = \{0, 25, 75, 130, 140, 255\}$	**0.796 (1)**
		$G = \{0, 30, 90, 170, 210, 255\}$	
		$B = \{0, 40, 80, 160, 200, 255\}$	
	2	$R = \{0, 80, 90, 190, 240, 255\}$	1.000 (2)
		$G = \{0, 60, 95, 185, 200, 255\}$	
		$B = \{0, 50, 100, 230, 235, 255\}$	

Table 4.7 Optimised and fixed class boundaries and evaluated segmentation quality measures, F' for 6 classes of Baboon image

EF	Set	Colour levels	η
F'_{op}	1	$R = \{0, 43, 170, 212, 233, 255\}$	**0.497 (1)**
		$G = \{0, 24, 79, 181, 201, 255\}$	
		$B = \{0, 54, 79, 194, 203, 255\}$	
	2	$R = \{0, 62, 124, 182, 184, 255\}$	0.533 (2)
		$G = \{0, 15, 44, 55, 132, 255\}$	
		$B = \{0, 5, 92, 122, 213, 255\}$	
F'_{fx}	1	$R = \{0, 21, 75, 115, 136, 255\}$	**0.632 (1)**
		$G = \{0, 24, 147, 177, 200, 255\}$	
		$B = \{0, 21, 62, 125, 202, 255\}$	
	2	$R = \{0, 70, 116, 151, 165, 255\}$	1.000 (2)
		$G = \{0, 41, 157, 180, 200, 255\}$	
		$B = \{0, 53, 88, 145, 162, 255\}$	

4.5.2.1 ParaOptiMUSIG Guided Segmented Outputs

The proposed ParaOptiMUSIG activation function based PSONN architecture is applied to generate the segmented multilevel test images for $K = 8$ and corresponding to the best segmentation quality measures (ρ, F, F', Q) achieved, images are shown in Figs. 4.4, 4.5, 4.6, 4.7, 4.8, 4.9, 4.10, 4.11, 4.12, 4.13, 4.14 and 4.15.

Table 4.8 Optimised and fixed class boundaries and evaluated segmentation quality measures, Q for 6 classes of Baboon image

EF	Set	Colour levels	η
Q_{op}	1	$R = \{0, 95, 110, 129, 174, 255\}$	0.655 (2)
		$G = \{0, 37, 72, 120, 181, 255\}$	
		$B = \{0, 36, 68, 84, 107, 255\}$	
	2	**$R = \{0, 78, 130, 149, 172, 255\}$**	**0.611 (1)**
		$G = \{0, 80, 92, 108, 147, 255\}$	
		$B = \{0, 37, 74, 80, 125, 255\}$	
Q_{fx}	1	**$R = \{0, 158, 170, 185, 221, 255\}$**	**0.859 (1)**
		$G = \{0, 66, 106, 159, 228, 255\}$	
		$B = \{0, 93, 124, 129, 139, 255\}$	
	2	$R = \{0, 36, 144, 201, 219, 255\}$	1.000 (2)
		$G = \{0, 54, 69, 146, 190, 255\}$	
		$B = \{0, 51, 117, 122, 151, 255\}$	

Table 4.9 Optimised and fixed class boundaries and evaluated segmentation quality measures, ρ for 8 classes of Lena image

EF	Set	Colour levels	η
ρ_{op}	1	**$R = \{43, 89, 156, 160, 172, 213, 237, 255\}$**	**0.9428 (1)**
		$G = \{0, 34, 67, 110, 151, 155, 187, 255\}$	
		$B = \{32, 67, 91, 107, 126, 164, 186, 238\}$	
	2	$R = \{43, 104, 129, 134, 155, 165, 183, 255\}$	0.9422 (2)
		$G = \{0, 13, 57, 147, 160, 207, 221, 255\}$	
		$B = \{32, 66, 89, 118, 143, 167, 190, 238\}$	
ρ_{fx}	1	$R = \{43, 80, 90, 110, 120, 190, 200, 255\}$	0.9184 (2)
		$G = \{0, 30, 75, 80, 100, 160, 220, 255\}$	
		$B = \{32, 50, 70, 90, 110, 150, 170, 238\}$	
	2	**$R = \{43, 70, 80, 110, 130, 180, 210, 255\}$**	**0.9289 (1)**
		$G = \{0, 35, 70, 90, 120, 160, 200, 255\}$	
		$B = \{32, 45, 77, 85, 115, 145, 175, 238\}$	

Table 4.10 Optimised and fixed class boundaries and evaluated segmentation quality measures, F for 8 classes of Lena image

EF	Set	Colour levels	η
F_{op}	1	$R = \{43, 81, 129, 133, 174, 203, 254, 255\}$	**0.676 (1)**
		$G = \{0, 12, 74, 128, 155, 209, 217, 255\}$	
		$B = \{32, 50, 60, 124, 170, 174, 206, 238\}$	
	2	$R = \{43, 90, 92, 107, 125, 238, 244, 255\}$	0.713 (2)
		$G = \{0, 4, 59, 121, 129, 145, 157, 255\}$	
		$B = \{32, 60, 108, 153, 175, 176, 206, 238\}$	
F_{fx}	1	$R = \{43, 70, 120, 130, 150, 170, 190, 255\}$	**0.866 (1)**
		$G = \{0, 20, 140, 180, 190, 210, 230, 255\}$	
		$B = \{32, 50, 80, 90, 140, 150, 160, 238\}$	
	2	$R = \{43, 90, 118, 126, 136, 161, 175, 255\}$	1.000 (2)
		$G = \{0, 51, 121, 167, 176, 200, 210, 255\}$	
		$B = \{32, 43, 84, 98, 138, 165, 172, 238\}$	

Table 4.11 Optimised and fixed class boundaries and evaluated segmentation quality measures, F' for 8 classes of Lena image

EF	Set	Colour levels	η
F'_{op}	1	$R = \{43, 90, 92, 107, 125, 238, 244, 255\}$	**0.046 (1)**
		$G = \{0, 18, 131, 134, 156, 171, 200, 255\}$	
		$B = \{32, 57, 122, 150, 176, 185, 193, 238\}$	
	2	$R = \{43, 73, 92, 114, 123, 166, 246, 255\}$	0.054 (2)
		$G = \{0, 93, 94, 148, 150, 212, 225, 255\}$	
		$B = \{32, 54, 77, 113, 139, 166, 191, 238\}$	
F'_{fx}	1	$R = \{43, 75, 95, 125, 155, 185, 215, 255\}$	1.000 (2)
		$G = \{0, 40, 70, 100, 130, 160, 200, 255\}$	
		$B = \{32, 60, 105, 135, 175, 195, 215, 238\}$	
	2	$R = \{43, 100, 128, 141, 163, 179, 190, 255\}$	0.101 (1)
		$G = \{0, 50, 70, 89, 145, 206, 210, 255\}$	
		$B = \{32, 53, 75, 106, 111, 121, 160, 238\}$	

Table 4.12 Optimised and fixed class boundaries and evaluated segmentation quality measures, Q for 8 classes of Lena image

EF	Set	Colour levels	η
Q_{op}	1	$R = \{43, 141, 166, 167, 202, 208, 220, 255\}$ $G = \{0, 36, 48, 64, 136, 147, 182, 255\}$ $B = \{32, 49, 79, 93, 109, 139, 206, 238\}$	0.601 (2)
	2	$R = \{\mathbf{43, 167, 173, 207, 209, 224, 241, 255}\}$ $G = \{\mathbf{0, 45, 48, 71, 83, 115, 158, 255}\}$ $B = \{\mathbf{32, 61, 89, 91, 124, 132, 157, 238}\}$	**0.581 (1)**
Q_{fx}	1	$R = \{\mathbf{43, 75, 95, 125, 155, 185, 215, 255}\}$ $G = \{\mathbf{0, 40, 70, 100, 130, 160, 200, 255}\}$ $B = \{\mathbf{32, 60, 105, 135, 175, 195, 215, 238}\}$	**0.833 (1)**
	2	$R = \{43, 45, 65, 85, 135, 175, 200, 255\}$ $G = \{0, 14, 62, 87, 137, 173, 210, 255\}$ $B = \{32, 41, 60, 82, 134, 168, 212, 238\}$	1.000 (2)

Table 4.13 Optimised and fixed class boundaries and evaluated segmentation quality measures, ρ for 8 classes of Baboon image

EF	Set	Colour levels	η
ρ_{op}	1	$R = \{0, 94, 106, 146, 173, 186, 203, 255\}$ $G = \{0, 39, 69, 74, 88, 111, 147, 255\}$ $B = \{0, 52, 75, 90, 127, 162, 219, 255\}$	0.9623 (2)
	2	$R = \{\mathbf{0, 17, 52, 73, 83, 116, 204, 255}\}$ $G = \{\mathbf{0, 27, 102, 152, 153, 180, 254, 255}\}$ $B = \{\mathbf{0, 48, 89, 117, 136, 177, 230, 255}\}$	**0.9632 (1)**
ρ_{fx}	1	$R = \{\mathbf{0, 50, 60, 90, 100, 110, 150, 255}\}$ $G = \{\mathbf{0, 60, 110, 160, 170, 185, 200, 255}\}$ $B = \{\mathbf{0, 10, 20, 60, 80, 150, 220, 255}\}$	**0.9371 (1)**
	2	$R = \{0, 41, 48, 90, 93, 112, 134, 255\}$ $G = \{0, 25, 63, 163, 166, 172, 200, 255\}$ $B = \{0, 16, 27, 66, 153, 245, 249, 255\}$	0.9135 (2)

Table 4.14 Optimised and fixed class boundaries and evaluated segmentation quality measures, F for 8 classes of Baboon image

EF	Set	Colour levels	η
F_{op}	1	**$R = \{0, 44, 47, 66, 88, 231, 239, 255\}$**	**0.065 (1)**
		$G = \{0, 11, 70, 116, 123, 137, 148, 255\}$	
		$B = \{0, 23, 97, 165, 199, 200, 224, 255\}$	
	2	$R = \{0, 32, 53, 197, 204, 246, 249, 255\}$	0.069 (2)
		$G = \{0, 31, 86, 94, 108, 118, 183, 255\}$	
		$B = \{0, 61, 129, 163, 164, 188, 234, 255\}$	
F_{fx}	1	**$R = \{0, 50, 60, 80, 100, 240, 250, 255\}$**	**0.827 (1)**
		$G = \{0, 15, 40, 100, 160, 170, 190, 255\}$	
		$B = \{0, 40, 50, 110, 115, 220, 230, 255\}$	
	2	$R = \{0, 70, 90, 125, 135, 155, 180, 255\}$	1.000 (2)
		$G = \{0, 10, 25, 85, 145, 180, 195, 255\}$	
		$B = \{0, 20, 40, 50, 95, 125, 205, 255\}$	

Table 4.15 Optimised and fixed class boundaries and evaluated segmentation quality measures, F' for 8 classes of Baboon image

EF	Set	Colour levels	η
F'_{op}	1	**$R = \{0, 17, 59, 62, 81, 103, 246, 255\}$**	**0.516 (1)**
		$G = \{0, 24, 83, 129, 136, 150, 161, 255\}$	
		$B = \{0, 39, 113, 181, 215, 216, 240, 255\}$	
	2	$R = \{0, 23, 83, 88, 101, 203, 241, 255\}$	0.587 (2)
		$G = \{0, 27, 29, 64, 99, 185, 192, 255\}$	
		$B = \{0, 29, 46, 141, 199, 203, 244, 255\}$	
F'_{fx}	1	**$R = \{0, 21, 75, 95, 125, 165, 190, 255\}$**	**0.089 (1)**
		$G = \{0, 24, 72, 97, 127, 163, 200, 255\}$	
		$B = \{0, 21, 70, 92, 124, 158, 202, 255\}$	
	2	$R = \{0, 20, 80, 85, 98, 180, 200, 255\}$	1.000 (2)
		$G = \{0, 18, 32, 102, 153, 170, 205, 255\}$	
		$B = \{0, 45, 50, 100, 120, 140, 190, 255\}$	

Table 4.16 Optimised and fixed class boundaries and evaluated segmentation quality measures, Q for 8 classes of Baboon image

EF	Set	Colour levels	η
Q_{op}	1	$R = \{0, 47, 73, 96, 116, 129, 177, 255\}$	0.575 (2)
		$G = \{0, 71, 96, 113, 134, 161, 189, 255\}$	
		$B = \{0, 54, 122, 124, 148, 162, 195, 255\}$	
	2	$R = \{0, 58, 108, 112, 125, 128, 219, 255\}$	**0.538 (1)**
		$G = \{0, 27, 46, 89, 146, 158, 165, 255\}$	
		$B = \{0, 47, 71, 91, 113, 124, 162, 255\}$	
Q_{fx}	1	$R = \{0, 50, 60, 90, 100, 110, 150, 255\}$	**0.787 (1)**
		$G = \{0, 60, 110, 160, 170, 185, 200, 255\}$	
		$B = \{0, 10, 20, 60, 80, 150, 220, 255\}$	
	2	$R = \{0, 21, 40, 62, 115, 138, 180, 255\}$	1.000 (2)
		$G = \{0, 22, 30, 104, 120, 140, 200, 255\}$	
		$B = \{0, 10, 30, 120, 200, 210, 220, 255\}$	

Table 4.17 Optimised and fixed class boundaries and evaluated segmentation quality measures, ρ for 8 classes of Brain MRI image

EF	Set	Colour levels	η
ρ_{op}	1	$R = \{0, 57, 69, 145, 152, 182, 234, 255\}$	0.9702 (2)
		$G = \{0, 42, 68, 101, 108, 161, 229, 255\}$	
		$B = \{0, 18, 44, 74, 110, 139, 191, 255\}$	
	2	$R = \{0, 88, 139, 142, 154, 222, 227, 255\}$	**0.9728 (1)**
		$G = \{0, 55, 66, 102, 189, 197, 213, 255\}$	
		$B = \{0, 15, 33, 82, 117, 159, 222, 255\}$	
ρ_{fx}	1	$R = \{0, 45, 70, 100, 120, 150, 180, 255\}$	**0.9371 (1)**
		$G = \{0, 30, 50, 90, 110, 180, 200, 255\}$	
		$B = \{0, 43, 65, 86, 106, 140, 175, 255\}$	
	2	$R = \{0, 35, 90, 100, 115, 130, 140, 255\}$	0.9135 (2)
		$G = \{0, 5, 74, 130, 155, 175, 200, 255\}$	
		$B = \{0, 50, 80, 120, 150, 175, 215, 255\}$	

Table 4.18 Optimised and fixed class boundaries and evaluated segmentation quality measures, F for 8 classes of Brain MRI image

EF	Set	Colour levels	η
F_{op}	1	$R = \{0, 8, 37, 104, 150, 203, 245, 255\}$	**0.655 (1)**
		$G = \{0, 20, 43, 69, 132, 227, 233, 255\}$	
		$B = \{0, 103, 136, 147, 185, 222, 230, 255\}$	
	2	$R = \{0, 6, 193, 228, 244, 250, 252, 255\}$	0.689 (2)
		$G = \{0, 25, 43, 75, 87, 126, 149, 255\}$	
		$B = \{0, 14, 133, 237, 244, 249, 252, 255\}$	
F_{fx}	1	$R = \{0, 20, 50, 100, 150, 200, 245, 255\}$	**0.862 (1)**
		$G = \{0, 20, 50, 100, 130, 210, 230, 255\}$	
		$B = \{0, 80, 120, 140, 180, 220, 230, 255\}$	
	2	$R = \{0, 25, 55, 105, 155, 205, 245, 255\}$	1.000 (2)
		$G = \{0, 30, 60, 110, 135, 215, 225, 255\}$	
		$B = \{0, 70, 115, 145, 175, 220, 235, 255\}$	

Table 4.19 Optimised and fixed class boundaries and evaluated segmentation quality measures, F' for 8 classes of Brain MRI image

EF	Set	Colour levels	η
F'_{op}	1	$R = \{0, 13, 23, 83, 190, 210, 236, 255\}$	**0.466 (1)**
		$G = \{0, 13, 26, 123, 180, 191, 212, 255\}$	
		$B = \{0, 9, 67, 107, 149, 189, 227, 255\}$	
	2	$R = \{0, 12, 32, 58, 88, 100, 158, 255\}$	0.562 (2)
		$G = \{0, 16, 28, 81, 94, 181, 192, 255\}$	
		$B = \{0, 38, 77, 115, 152, 203, 209, 255\}$	
F'_{fx}	1	$R = \{0, 60, 90, 120, 150, 180, 250, 255\}$	**0.991 (1)**
		$G = \{0, 30, 80, 140, 170, 185, 210, 255\}$	
		$B = \{0, 50, 130, 160, 186, 195, 210, 255\}$	
	2	$R = \{0, 55, 85, 115, 145, 180, 225, 255\}$	1.000 (2)
		$G = \{0, 35, 80, 135, 165, 185, 215, 255\}$	
		$B = \{0, 50, 130, 160, 186, 195, 225, 255\}$	

Table 4.20 Optimised and fixed class boundaries and evaluated segmentation quality measures, Q for 8 classes of Brain MRI image

EF	Set	Colour levels	η
Q_{op}	1	$R = \{0, 47, 93, 109, 129, 178, 183, 255\}$	0.139 (2)
		$G = \{0, 32, 43, 97, 110, 207, 218, 255\}$	
		$B = \{0, 52, 137, 140, 152, 229, 252, 255\}$	
	2	$R = \{0, 18, 38, 51, 158, 214, 234, 255\}$	**0.114 (1)**
		$G = \{0, 40, 62, 88, 96, 172, 176, 255\}$	
		$B = \{0, 31, 120, 127, 174, 195, 203, 255\}$	
Q_{fx}	1	$R = \{0, 30, 60, 90, 140, 180, 205, 255\}$	**0.535 (1)**
		$G = \{0, 20, 50, 80, 135, 175, 215, 255\}$	
		$B = \{0, 35, 55, 75, 125, 155, 210, 255\}$	
	2	$R = \{0, 45, 65, 85, 135, 175, 200, 255\}$	1.000 (2)
		$G = \{0, 10, 50, 90, 130, 170, 210, 255\}$	
		$B = \{0, 40, 60, 75, 140, 150, 200, 255\}$	

(a)   **(b)**

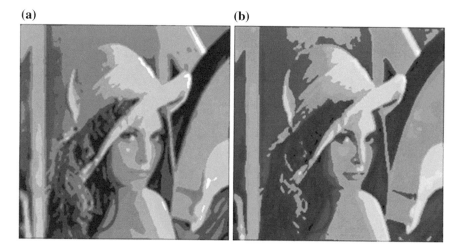

Fig. 4.4 8-class segmented 256×256 Lena image with the optimised class levels pertaining to **a** set 1 **b** set 2 of Table 4.9 for the quality measure ρ with ParaOptiMUSIG activation function

(a)   (b)

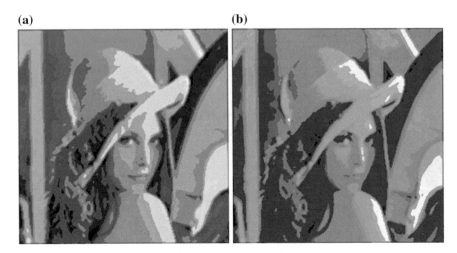

Fig. 4.5 8-class segmented 256×256 Lena image with the optimised class levels pertaining to **a** set 1 **b** set 2 of Table 4.10 for the quality measure F with ParaOptiMUSIG activation function

(a)   (b)

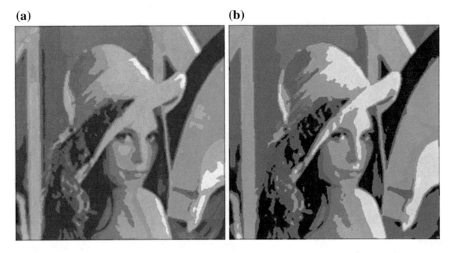

Fig. 4.6 8-class segmented 256×256 Lena image with the optimised class levels pertaining to **a** set 1 **b** set 2 of Table 4.11 for the quality measure F' with ParaOptiMUSIG activation function

4.5.2.2   MUSIG Guided Segmented Outputs

The segmented multicolour test images obtained with the PSONN architecture characterised by the conventional MUSIG activation employing fixed class responses for $K = 8$ and with four segmentation quality measures (ρ, F, F', Q) achieved, are shown in Figs. 4.16, 4.17, 4.18, 4.19, 4.20, 4.21, 4.22, 4.23, 4.24, 4.25, 4.26 and 4.27.

(a) **(b)**

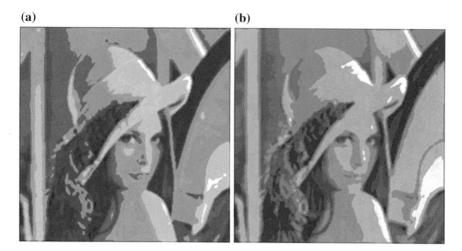

Fig. 4.7 8-class segmented 256 × 256 Lena image with the optimised class levels pertaining to **a** set 1 **b** set 2 of Table 4.12 for the quality measure Q with ParaOptiMUSIG activation function

(a) **(b)**

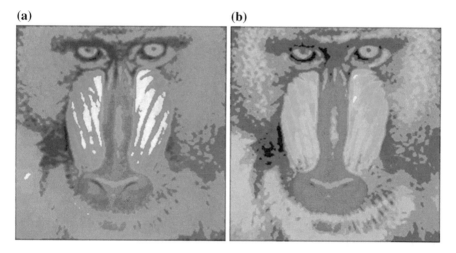

Fig. 4.8 8-class segmented 256 × 256 Baboon image with the optimised class levels pertaining to **a** set 1 **b** set 2 of Table 4.13 for the quality measure ρ with ParaOptiMUSIG activation function

A detailed analysis is presented on the basis of the reported segmentation quality measures and the reported results are tabulated in the Tables 4.1, 4.2, 4.3, 4.4, 4.5, 4.6, 4.7, 4.8, 4.9, 4.10, 4.11, 4.12, 4.13, 4.14, 4.15, 4.16, 4.17, 4.18, 4.19 and 4.20. Results, the optimised and fixed class boundaries and evaluated segmentation quality measures, for 6 classes are presented in Tables 4.1, 4.2, 4.3, 4.4, 4.5, 4.6, 4.7 and 4.8 and among them, the results for the Lena image are shown in the first four tables (Tables 4.1, 4.2, 4.3 and 4.4) and the remaining four tables (Tables 4.5, 4.6, 4.7 and 4.8) are expended to tabulate the results for the baboon image. In Table 4.1, both

(a)   **(b)**

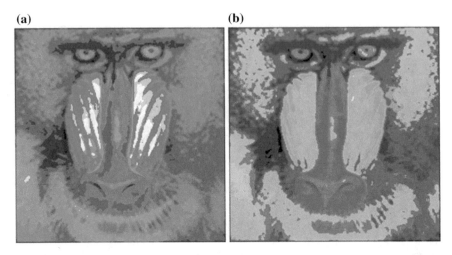

Fig. 4.9 8-class segmented 256 × 256 Baboon image with the optimised class levels pertaining to **a** set 1 **b** set 2 of Table 4.14 for the quality measure F with ParaOptiMUSIG activation function

(a)   **(b)**

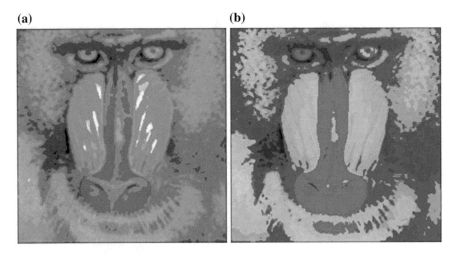

Fig. 4.10 8-class segmented 256 × 256 Baboon image with the optimised class levels pertaining to **a** set 1 **b** set 2 of Table 4.15 for the quality measure F' with ParaOptiMUSIG activation function

the ρ results, derived by the proposed ParaOptiMUSIG activation function are better than the same derived by the conventional MUSIG activation function. In the next Table, the ParaOptiMUSIG activation function based F values for the Lena image are far better than the MUSIG activation function based F values. The optimised and fixed class boundaries and evaluated segmentation quality measures, F' and Q for 6 classes of Lena image are tabulated in Tables 4.3 and 4.4, respectively. The results derived by the proposed method are better than the conventional MUSIG activation function in both the tables. In the next four tables (Tables 4.5, 4.6, 4.7 and

(a)   **(b)**

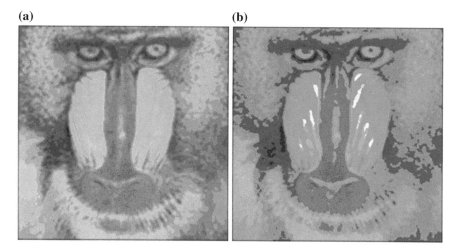

Fig. 4.11 8-class segmented 256×256 Baboon image with the optimised class levels pertaining to **a** set 1 **b** set 2 of Table 4.16 for the quality measure Q with ParaOptiMUSIG activation function

(a)   **(b)**

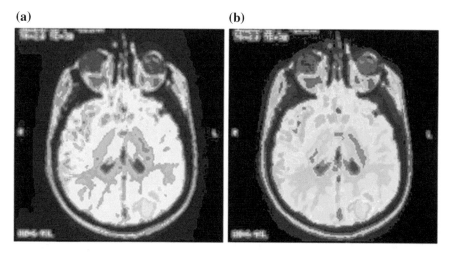

Fig. 4.12 8-class segmented 170×170 Brain MRI image with the optimised class levels pertaining to **a** set 1 **b** set 2 of Table 4.17 for the quality measure ρ with ParaOptiMUSIG activation function

4.8), the optimised and fixed class boundaries and evaluated segmentation quality measures are reported for Baboon image and that image is segmented in six classes. The derived ρ, F, F', and Q values in those four tables by the proposed method are doing well than the same values obtained by the MUSIG activation function. The same experiment is again performed by segmenting the test images in 8 classes and now the experiment is done on three test images. In Tables 4.9, 4.13 and 4.17 for Lena, Baboon and Brain MRI images, respectively, the optimised and fixed class boundaries and evaluated segmentation quality measures, ρ, for 8 classes are tabulated and it

(a)   **(b)**

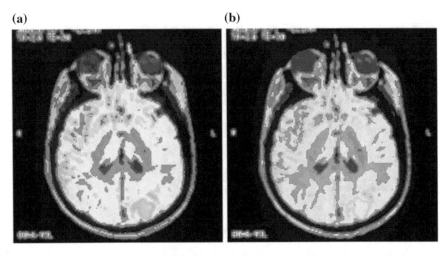

Fig. 4.13 8-class segmented 170×170 Brain MRI image with the optimised class levels pertaining to **a** set 1 **b** set 2 of Table 4.18 for the quality measure F with ParaOptiMUSIG activation function

(a)   **(b)**

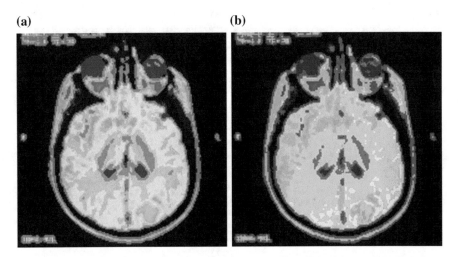

Fig. 4.14 8-class segmented 170×170 Brain MRI image with the optimised class levels pertaining to **a** set 1 **b** set 2 of Table 4.19 for the quality measure F' with ParaOptiMUSIG activation function

is easy to understand that the ρ values derived by the proposed method is much more encouraging than the same evaluated by the conventional MUSIG activation function. In Tables 4.10, 4.14 and 4.17 for different images, the F values gained by the ParaOptiMUSIG activation function are smaller than the same deduced by the MUSIG activation function and we can say that the proposed method is good as smaller F value indicates better segmentation. Similarly, the smaller F' values are gained by the proposed method than the same deduced by the MUSIG activation

(a)   (b)

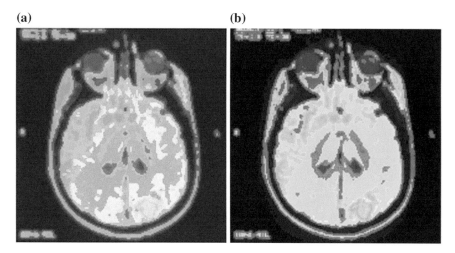

Fig. 4.15 8-class segmented 170×170 Brain MRI image with the optimised class levels pertaining to **a** set 1 **b** set 2 of Table 4.20 for the quality measure Q with ParaOptiMUSIG activation function

(a)   (b)

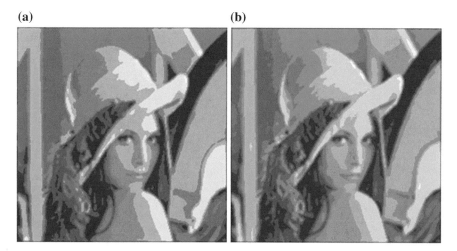

Fig. 4.16 8-class segmented 256×256 Lena image with the fixed class levels pertaining to **a** set 1 **b** set 2 of Table 4.9 for the quality measure ρ with MUSIG activation function

function and the corresponding results are tabulated in Tables 4.11, 4.15 and 4.19 for three test images. The segmentation quality measure, Q is determined for three test images and they are tabulated in Tables 4.12, 4.16 and 4.20. It is observed that the proposed method generates better Q values than the conventional MUSIG activation function.

It is quite clear from the derived results that the ParaOptiMUSIG activation function outperforms its conventional MUSIG counterpart in respect of the segmentation quality of the images for the different number of classes. It is also evident that the

(a) (b)

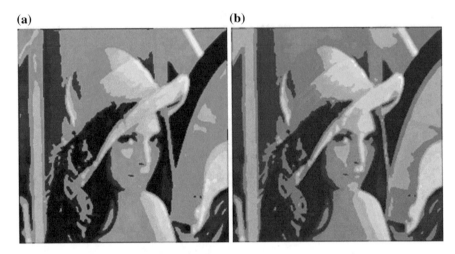

Fig. 4.17 8-class segmented 256 × 256 Lena image with the fixed class levels pertaining to **a** set 1 **b** set 2 of Table 4.10 for the quality measure F with MUSIG activation function

(a) (b)

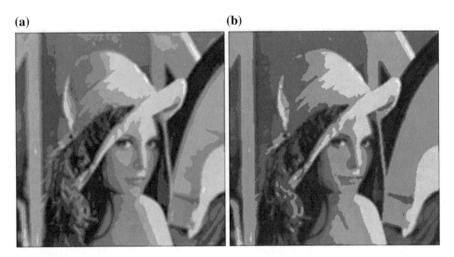

Fig. 4.18 8-class segmented 256 × 256 Lena image with the fixed class levels pertaining to **a** set 1 **b** set 2 of Table 4.11 for the quality measure F' with MUSIG activation function

ParaOptiMUSIG function incorporates the image heterogeneity as it can handle a wide variety of image intensity distribution prevalent in real life.

(a)   **(b)**

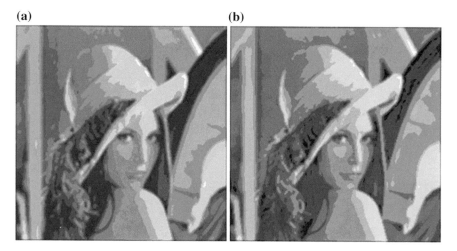

Fig. 4.19 8-class segmented 256 × 256 Lena image with the fixed class levels pertaining to **a** set 1 **b** set 2 of Table 4.12 for the quality measure Q with MUSIG activation function

(a)   **(b)**

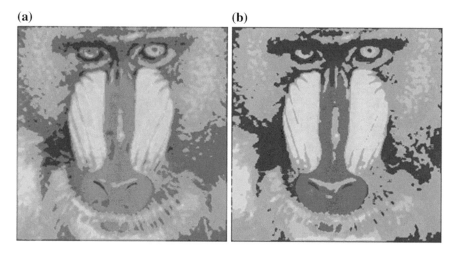

Fig. 4.20 8-class segmented 256 × 256 Baboon image with the fixed class levels pertaining to **a** set 1 **b** set 2 of Table 4.13 for the quality measure ρ with MUSIG activation function

4.6 Discussions and Conclusion

A segmentation procedure of colour images has been demonstrated with the parallel self-organising neural network (PSONN) architecture in this chapter. A parallel optimised multilevel MUSIG activation function is described to introduce the multiscaling capabilities in the network. In this chapter, it has been noted that the fixed and heuristic class responses are employed in the MUSIG activation function. The

(a)   **(b)**

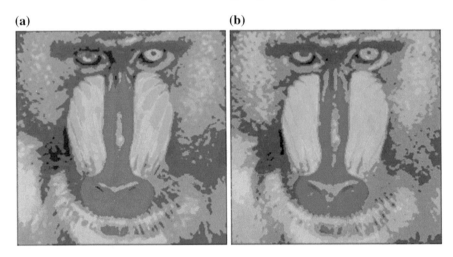

Fig. 4.21 8-class segmented 256 × 256 Baboon image with the fixed class levels pertaining to **a** set 1 **b** set 2 of Table 4.14 for the quality measure F with MUSIG activation function

(a)   **(b)**

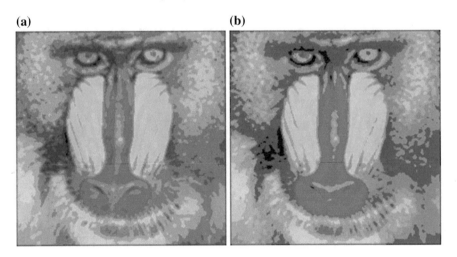

Fig. 4.22 8-class segmented 256 × 256 Baboon image with the fixed class levels pertaining to **a** set 1 **b** set 2 of Table 4.15 for the quality measure F' with MUSIG activation function

MUSIG activation function is unaware of the underneath information of the image data. A parallel version of the optimised MUSIG (ParaOptiMUSIG) activation function is purported with the optimised class boundaries by incorporating the intensity gamut immanent in the input images. The optimised class boundaries are deduced by a genetic algorithm based optimization procedure with three different entropy-based objective functions. These functions are treated as measures of image segmentation quality. The performance of the proposed ParaOptiMUSIG activation function for the segmentation of real-life true colour images indicate superior performance as

(a)   **(b)**

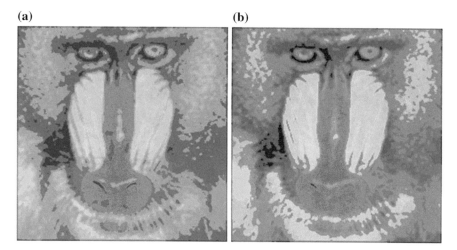

Fig. 4.23 8-class segmented 256 × 256 Baboon image with the fixed class levels pertaining to **a** set 1 **b** set 2 of Table 4.16 for the quality measure Q with MUSIG activation function

(a)   **(b)**

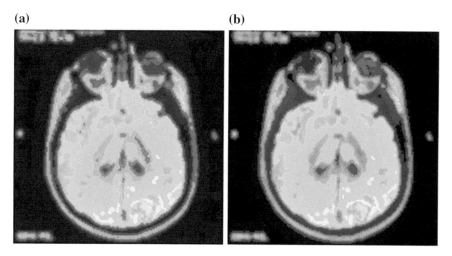

Fig. 4.24 8-class segmented 170 × 170 Brain MRI image with the fixed class levels pertaining to **a** set 1 **b** set 2 of Table 4.17 for the quality measure ρ with MUSIG activation function

compared to the conventional MUSIG activation function with fixed and heuristic class levels. Application of the proposed activation function with modified functionalities is also demonstrated for the segmentation of true colour images.

The class boundaries are generated on the basis of the single objective function by the proposed method. These class boundaries are applied to generate the proposed activation functions. It may not give any guarantee that a set of evolved class boundaries for one objective function will give good result for another objective function. The derived class boundaries by the proposed methods are objective function

(a)   **(b)**

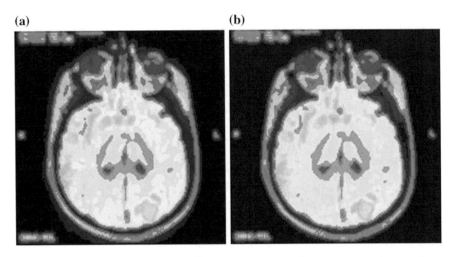

Fig. 4.25 8-class segmented 170×170 Brain MRI image with the fixed class levels pertaining to **a** set 1 **b** set 2 of Table 4.18 for the quality measure F with MUSIG activation function

(a)   **(b)**

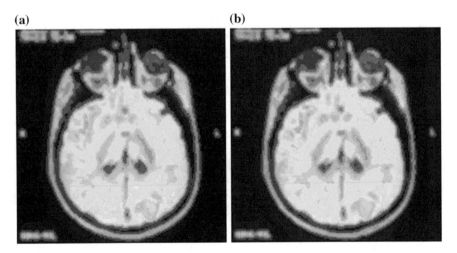

Fig. 4.26 8-class segmented 170×170 Brain MRI image with the fixed class levels pertaining to **a** set 1 **b** set 2 of Table 4.19 for the quality measure F' with MUSIG activation function

dependent. To solve this kind of problem, we look after a solution which will work for multi-objective criteria. This means that the class boundaries have to be generated on the basis of multi-objective functions and the derived class boundaries will be used to generate the activation functions. This is true for the proposed OptiMUSIG

(a)   **(b)**

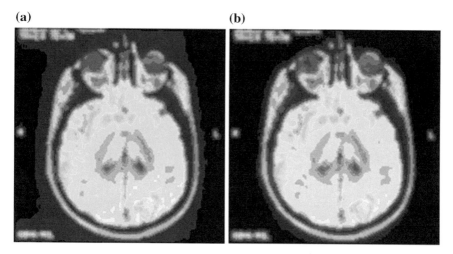

Fig. 4.27 8-class segmented 170×170 Brain MRI image with the fixed class levels pertaining to **a** set 1 **b** set 2 of Table 4.20 for the quality measure Q with MUSIG activation function

activation function which is discussed in the previous chapter and also for the proposed ParaOptiMUSIG activation function which is presented in this chapter. The following chapters of this book are aimed in this direction.

Chapter 5
Self-supervised Grey Level Image Segmentation Using Multi-Objective-Based Optimised MUSIG (OptiMUSIG) Activation Function

5.1 Introduction

The multilevel greyscale image can be efficiently be segmented by the OptiMUSIG activation function with the help of the multilayer self-organizing neural network (MLSONN) architecture. This activation function is more capable to incorporate the heterogeneous information content of the test images than the MUSIG activation function. The working principle and functionality of this activation function is already discussed in Chap. 3. One thing we have to keep in mind that the class levels which are used to generate the activation function are derived on the basis of single objective criterion. This method may or may not generate a good quality segmented image as the segmentation criteria of these methods are based on a single objective or single segmentation evaluation criterion. In other words, the derived class levels on the basis of a particular objective function may not be good one for another objective function.

In order to solve a certain problem in many real-world situations, several incommensurable and competing constraints of that problem have to be optimised simultaneously. Usually, there will be a set of alternative solutions instead of a single optimal solution. These solutions are considered as optimal solutions so that no other solutions are superior to these solutions when all constraints are considered. The foremost problem considering multi-objective optimisation (MOO) is that there is no accepted definition of optimum in this case and therefore it is difficult to compare one solution with another one. These set of optimal solutions is defined as the Pareto optimal solutions.

Image segmentation is an important real-world problem and several segmentation algorithms are usually applied to optimise some evaluation/goodness measures. Simultaneous optimisation of multiple segmentation evaluation measures serves to handle with different characteristics of segmentation and generates good segmented images when all constraints are considered. In these type of multi-objective optimisation (MOO) [98] problems, searching of solutions is executed over a number of,

© Springer International Publishing AG 2016
S. De et al., *Hybrid Soft Computing for Multilevel Image and Data Segmentation*, Computational Intelligence Methods and Applications,
DOI 10.1007/978-3-319-47524-0_5

often conflicting, objective functions. Instead of getting a single optimal solution, a set of optimal solutions are generated and it is difficult to compare one solution with another one. These set of optimal solutions is defined as the Pareto optimal solutions. Motivated by this, this article depicts a multi-objective neuro-genetic image segmentation technique that optimises some evaluation/goodness measures simultaneously. In the first portion of this chapter a multi-objective genetic algorithm (MOGA) based approach has been presented and after that, a popular evolutionary multi-objective optimisation technique, named Non-dominated Sorting Genetic Algorithm-II (NSGA-II) [98, 101, 102] has been applied in this presented approach as the underlying optimisation technique.

The multi-objective optimisation (MOO) [98] methods can optimise multiple cluster validity measures simultaneously to cope with different characteristics of the partitioning and leads to higher quality solutions. Several modern techniques for MOO problems are used by the researchers. Mainly, the GA-based optimisation techniques such as multi-objective genetic algorithm (MOGA) [98], non-dominated sorting genetic algorithm (NSGA) [98, 112], NSGA-II [98, 101, 102], strength pareto evolutionary algorithm (SPEA) [113], SPEA 2 [114] are very popular. A detailed functionality and the working procedure of multi-objective optimisation using evolutionary algorithms are reported in different articles [276, 277] and those articles can arise interest to the aspirant researchers. There are several approaches to solve the problem of image segmentation by the previously mentioned multi-objective genetic algorithms. Several multi-objective optimisation techniques for image segmentation can be found in the literature [278]. Bandyopadhyay et al. [103] presented a multi-objective genetic clustering-based method to segregate the pixels of remote sensing images. NSGA-II [98, 101, 102] has been applied in this method to handle the problem of fuzzy partitioning by optimising different fuzzy clustering indices simultaneously. Mukhopadhyay and Maulik [106] proposed a multi-objective real-coded genetic fuzzy clustering method to segment the remote sensing images. In this method, NSGA-II is applied as the multi-objective optimisation technique in accordance with two conflicting fitness functions. Two important factors of the image segmentation, overall deviation and edge value, can be optimised simultaneously using a multi-objective evolutionary algorithm [279]. An improved version of the NSGA-II has been presented by Nakib et al. [280] to solve the segmentation problem by changing the stopping criterion and automating the choice of the size of the initial population and the number of generations.

This chapter is aimed at providing a bird's eye view to overcome the drawback of the single objective-based OptiMUSIG activation function for segmentation of multilevel greyscale images. Keeping this concept in view, in the first part of this chapter, the multilevel greyscale images are segmented into different number of classes with the multi-objective genetic algorithm-based optimised class levels. The resultant multi-objective genetic algorithm-based class levels are applied to generate an optimised MUSIG (OptiMUSIG) activation function for producing multilevel greyscale image segmentation using a single MLSONN architecture. Two real-life multilevel greyscale images, viz. the Lena and Baboon images, and a multilevel biomedical images of brain, viz. a Brain Neuroanatomy (MRI) image [253] have been applied to demonstrate the applications of the multi-objective genetic algorithm-based OptiMUSIG activation function approach. The multi-objective genetic algorithm is employed to optimise three measures, viz., the standard measure of correlation coefficient (ρ) [72], F due to Liu and Yang [255] and F' due to Borsotti et al. [256]. Results of segmentation using the multi-objective genetic algorithm-based OptiMUSIG shows better performance over the conventional MUSIG activation function employing heuristic class responses. In the second part of this chapter, the multilevel greyscale images are segmented into different number of classes with the NSGA-II-based optimised class levels which are employed to generate an optimised MUSIG (OptiMUSIG) activation function. This activation function is applied in a single MLSONN architecture for segmenting multilevel greyscale images. The same three fitness functions are used in this NSGA-II-based method. Ultimately, it has been proved on the basis of the results that the segmentation using the NSGA-II-based OptiMUSIG perform better than the conventional MUSIG activation function employing heuristic class responses. In both the cases, another Q index [256] is applied to select the particular solution from the Pareto optimal set of solutions and the selected solution is applied to generate the activation function. After that the generated activation function is applied in the MLSONN network architecture to segment the multilevel greyscale images. This approach is illustrated with two real-life multilevel greyscale images, viz. the Baboon and Peppers images, and a multilevel biomedical images of brain, viz. a Brain Neuroanatomy (MRI) image [253].

The outline of this chapter is as follows. The next section discusses the detailed representation of the multi-objective genetic algorithm-based methodology to segment the multilevel greyscale images. The image segmentation results using this method, both in qualitative and quantitative nature, are detailed in the next section. The NSGA-II-based OptiMUSIG activation function is in the next section and this section is followed by the multilevel greyscale image segmentation procedure by the NSGA-II-based OptiMUSIG activation function. Both the qualitative and quantitative comparisons are demonstrated in this section in between the image segmentation by the NSGA-II-based OptiMUSIG activation function-based method and the MUSIG activation function. The chapter ends with a brief concluding section.

5.2 Multilevel Greyscale Image Segmentation by Multi-objective Genetic Algorithm-Based OptiMUSIG Activation Function

The multilevel greyscale images are efficiently segmented by the OptiMUSIG activation function which has been discussed and proved in Chap. 3. The optimised class levels, generated by that approach in that chapter, are derived on the basis of single objective functions. It has been quite clear from the previous discussion that the single objective-based OptiMUSIG activation function may not generate good multilevel greyscale segmented output image in respect of all other fitness criteria.

Fig. 5.1 Flow diagram of greyscale image segmentation using multi-objective genetic algorithm-based OptiMUSIG activation function

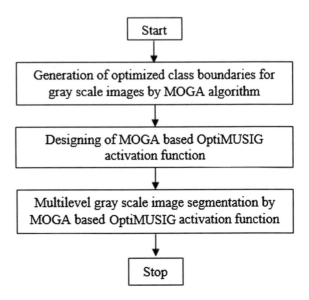

In this chapter, the optimised class levels are generated on the basis of more than one criterion to overcome the problem. The derived class levels are efficient for all functional objectives.

The multilevel greyscale image segmentation by a multi-objective genetic algorithm-based OptiMUSIG activation function with a MLSONN architecture has been presented in this section and the flow diagram of this approach is depicted in Fig. 5.1. This section is categorised into two subsections. The methodology is described in the first part of this section and the second part of this section shows the experimental result of that methodology [99].

5.2.1 Methodology

The first part of this chapter focuses to segment the greyscale image by the multi-objective-based OptiMUSIG activation function with a MLSONN architecture. This approach has been carried out into three phases. The different phases are discussed in the following subsections [99].

5.2.1.1 Generation of Optimised Class Boundaries for Greyscale Images

In this most important phase of the approach, the multi-objective genetic algorithm-based optimisation procedure is applied to generate the optimised class boundaries

$(c_{\gamma_{opt}})$ of the presented multi-objective genetic algorithm-based OptiMUSIG activation function. The technique is described below in detail.

- *Initialization phase*: The number of classes (K) to be segmented and the pixel intensity levels of the greyscale image are provided as inputs to the multi-objective genetic algorithm-based optimisation procedure in this phase.
- *Chromosome representation and population generation*: A binary encoding technique for the chromosomes is used for developing the optimised class boundaries from the input greyscale image information content. Randomly selected binary combinations of eight bits represent the class boundary level of the segmentation. If there are K segments in the image, the size of the chromosome depends on the $K \times 8$ bits and the class boundaries encoded in a chromosome in the initial population are randomly chosen to obtain K distinct class boundaries from the image content. A population size of 200 has been applied for this treatment.
- *Fitness computation*: In this phase, three segmentation efficiency measures (ρ, F, F') given in Eqs. 3.25, 3.28 and 3.29 respectively, are applied as the fitness function. These functions are used as the evaluation functions in the multi-objective genetic algorithm.
- *Selection*: The non-dominated solutions are selected from the chromosome pool in this phase. These selected chromosomes are applied to generate the mating pool for the crossover and mutation.
- *Crossover*: A single point crossover operation is applied to generate the new pool of chromosomes. In this approach, the crossover probability is set to 0.8.
- *Mutation*: The mutation probability is taken as 0.01 in this approach.
- *New generation*: After mutation, the child chromosomes are combined with previously selected chromosomes to make the new generation. This new population is again propagated to form the new generation.

This process is continued for a certain number of iterations. In this approach, this process proceeded for 100 iterations. After this process, the Pareto optimal set of chromosomes are evolved.

- *Selecting a solution from the non-dominated set*: It is necessary to choose a particular solution from the set of non-dominated solutions and this can be determined by a suitable cluster validity index. There are several cluster validity indices reported in different literatures, like Davies-Bouldin (DB) index [281], CDbw (Composed Density between and within clusters) [282]. However, it does not signify that every clustering index can be applied in every sector of clustering application. The best number of clustering in remotely sensed images can be correctly evaluated by the DB index [170, 283] and it is noted that the DB index was utilised in combining with another index in the unsupervised classification method [170, 284]. However, DB index [281] is very much noise sensitive and it works efficiently only for spherical clusters [285]. Hence, the image clustering results can be assessed by the image segmentation evaluation index like entropy-based index [286] or quantitative-based index [256]. A lower quantitative value or entropy value leads to better segmentations. The set of class levels that produces a set of clusters which minimises the Q index [256] exhibits a good clustering. The better chromosomes, having the minimum Q value, are applied to generate the OptiMUSIG activation function.

5.2.1.2   Designing of Multi-objective Genetic Algorithm-Based OptiMUSIG Activation Function

The optimised class boundaries ($c_{\gamma_{opt}}$) of the selected chromosomes from the Pareto optimal non-dominated set are employed to design the OptiMUSIG activation function. The $\alpha_{\gamma_{opt}}$ parameters are determined using the optimised $c_{\gamma_{opt}}$ by the Eq. 3.23. These $\alpha_{\gamma_{opt}}$ parameters are further employed to obtain the different transition levels of the OptiMUSIG activation function.

5.2.1.3   Multilevel Greyscale Image Segmentation by the Multi-objective Genetic Algorithm-Based OptiMUSIG Activation Function

In the ultimate phase of this approach, a single MLSONN architecture characterized by the multi-objective genetic algorithm-based OptiMUSIG activation function is utilised to segment the real-life multilevel greyscale images. The processed input information propagates to the succeeding network layers. Different grey level responses to the input image information have been evaluated by the neurons of the different layers of the MLSONN architecture. The MLSONN architecture has no *a priori* knowledge about the outputs as it is a self-supervised neural network. Thus, the system errors are determined by the subnormal linear index of fuzziness [72] at the output layer of the MLSONN architecture. These errors are used to adjust the interconnection weights between the different layers using the standard backpropagation algorithm [37, 47]. The system errors are minimised by transferring the outputs of the output layer of the network to the input layer of the network for further processing. The original input image gets segmented into different multilevel regions depending upon the optimised transition levels of the OptiMUSIG activation function after the self-supervision of the network attains stabilisation.

5.2.2   Experimental Results

The multilevel greyscale image segmentation approach using the MLSONN architecture guided by the multi-objective genetic algorithm-based OptiMUSIG activation function is demonstrated with two real-life images, viz. Lena and Baboon each of dimensions 128×128 and a multilevel biomedical images of brain, viz. a Brain Neuroanatomy (MRI) image [253]. Experiments have been conducted with $K = \{6, 8\}$ classes. The OptiMUSIG activation function has been designed with a fixed slope, $\lambda = \{2, 4\}$. Results are reported for eight classes and a fixed slope of $\lambda = 4$. The segmentation efficiency of this approach in accordance with the multi-objective genetic algorithm-based OptiMUSIG is shown in the following subsection. The narrated approach has been compared with the segmentation derived by means of the conventional MUSIG activation function with same number of class responses and heuristic class levels. The quantitative comparison of the above-mentioned methods is demonstrated in this subsection. In the next subsection, the comparative study as regards to its efficacy in the segmentation of test images of the segmented outputs by the narrated methods and the conventional MUSIG activation function-based methods is elaborated.

5.2.2.1 Quantitative Performance Analysis of Segmentation

The quantitative measures of the efficiency of the multi-objective genetic algorithm-based OptiMUSIG and the conventional MUSIG activation functions for $K = 8$ have been illustrated in this section. Three segmentation efficiency functions, viz. correlation coefficient (ρ), and two empirical evaluation functions (F and F') have been applied in the multi-objective genetic algorithm-based approaches to generate the activation function and another empirical evaluation function (Q) has been used for the selection of the chromosome that is applied in the corresponding network in the recounted method. The results obtained with the multi-objective genetic algorithm-based OptiMUSIG activation function have been discussed in the following subsection and the corresponding results obtained with the conventional fixed class response based MUSIG activation function are furnished in the next subsection.

1. *Segmentation evaluation using the multi-objective genetic algorithm-based Opti-MUSIG activation function for greyscale images*: The optimised set of class boundaries ($c_{\gamma_{opt}}$) for the greyscale images obtained using multi-objective genetic algorithm-based optimisation with the three evaluation functions (ρ, F and F') are shown in Tables 5.1, 5.2 and 5.3. The evaluation functions are treated as the fitness functions in the multi-objective genetic algorithm-based procedure. The corresponding three evaluation functions (ρ, F and F') values of individual chromosome set are also reported in these tables. The optimised set of class boundaries are derived as the Pareto optimal set by the narrated approach. Four Pareto optimal sets are reported for Lena and Baboon images and two Pareto optimal sets are tabulated for the Brain MRI image. The Q values of individual set of class boundaries for greyscale images are presented in Tables 5.4, 5.5 and 5.6. The last columns of the Tables 5.1, 5.2 and 5.3 show the quality measures κ [graded on a scale of 1 (best) onwards] obtained by the segmentation of the test images based on the corresponding set of optimised class boundaries. The gradation has been performed on the basis of the Q values of the particular set of class boundaries. The first four good Q values obtained are indicated in those tables for easy reckoning. All evaluation function values are tabulated in normalised form in this chapter.

Table 5.1 Pareto optimal set of optimised class boundaries and corresponding evaluated segmentation quality measures for eight classes of greyscale Lena image

Set no		Class level	ρ	F	F'	κ
1	(i)	0, 50, 84, 119, 146, 172, 206, 255	0.9800	0.6941	0.6955	
	(ii)	0, 50, 84, 117, 146, 172, 206, 255	0.9797	0.6931	0.6944	
	(iii)	0, 73, 98, 123, 148, 177, 205, 255	0.9591	0.6585	0.6585	3
	(iv)	0, 50, 88, 118, 146, 172, 206, 255	0.9807	0.6994	0.7008	
	(v)	0, 73, 98, 125, 148, 177, 205, 255	0.9591	0.6597	0.6596	1
	(vi)	0, 73, 98, 119, 144, 177, 205, 255	0.9579	0.6375	0.6374	2
	(vii)	0, 73, 102, 117, 144, 177, 205, 255	0.9568	0.6288	0.6287	
	(viii)	0, 50, 88, 119, 146, 175, 206, 255	0.9808	0.7041	0.7055	
	(ix)	0, 73, 98, 117, 144, 177, 205, 255	0.9577	0.6317	0.6316	4
	(x)	0, 73, 98, 119, 146, 177, 205, 255	0.9580	0.6482	0.6482	
	(xi)	0, 50, 88, 119, 146, 171, 206, 255	0.9811	0.7113	0.7128	
2	(i)	0, 53, 89, 122, 151, 175, 205, 255	0.9810	0.8164	0.8164	
	(ii)	0, 68, 113, 147, 169, 185, 205, 255	0.9556	0.6339	0.6339	
	(iii)	0, 47, 84, 119, 147, 173, 205, 255	0.9800	0.7467	0.7469	
	(iv)	0, 49, 89, 119, 147, 173, 205, 255	0.9810	0.7546	0.7546	2
	(v)	0, 47, 80, 117, 147, 171, 205, 255	0.9785	0.7260	0.7260	
	(vi)	0, 47, 84, 117, 147, 173, 205, 255	0.9797	0.7438	0.7438	
	(vii)	0, 47, 82, 103, 147, 171, 205, 255	0.9747	0.7040	0.7039	
	(viii)	0, 49, 88, 117, 147, 173, 205, 255	0.9803	0.7511	0.7511	4
	(ix)	0, 47, 82, 117, 147, 171, 205, 255	0.9792	0.7273	0.7273	
	(x)	0, 47, 84, 116, 147, 173, 205, 255	0.9797	0.7438	0.7438	
	(xi)	0, 72, 103, 120, 146, 175, 204, 255	0.9585	0.6468	0.6468	1
	(xii)	0, 49, 89, 118, 147, 173, 205, 255	0.9810	0.7526	0.7526	3
	(xiii)	0, 47, 74, 117, 147, 171, 205, 255	0.9758	0.7163	0.7163	
3	(i)	0, 50, 76, 110, 144, 174, 205, 255	0.9772	0.7070	0.7083	
	(ii)	0, 69, 105, 120, 137, 183, 205, 255	0.9535	0.5934	0.5934	3
	(iii)	0, 48, 89, 119, 147, 172, 205, 255	0.9806	0.7529	0.7529	
	(iv)	0, 50, 76, 110, 144, 172, 205, 255	0.9776	0.7185	0.7199	
	(v)	0, 70, 102, 121, 144, 174, 207, 255	0.9607	0.6026	0.6026	2
	(vi)	0, 50, 78, 110, 144, 174, 205, 255	0.9776	0.7089	0.7102	
	(vii)	0, 70, 102, 125, 144, 174, 207, 255	0.9609	0.6079	0.6079	1
	(viii)	0, 50, 70, 110, 144, 174, 207, 255	0.9754	0.6995	0.7008	4
	(ix)	0, 48, 89, 117, 147, 174, 207, 255	0.9801	0.7526	0.7526	
	(x)	0, 50, 78, 106, 144, 174, 207, 255	0.9771	0.7035	0.7048	
	(xi)	0, 50, 70, 103, 144, 174, 207, 255	0.9753	0.6954	0.6967	

(continued)

Table 5.1 (continued)

Set no		Class level	ρ	F	F'	κ
4	(i)	0, 53, 120, 136, 157, 177, 205, 255	0.9595	0.6507	0.6507	
	(ii)	0, 49, 86, 112, 144, 174, 207, 255	0.9792	0.7340	0.7339	
	(iii)	0, 49, 77, 112, 144, 174, 207, 255	0.9774	0.7124	0.7124	
	(iv)	0, 74, 99, 121, 146, 172, 206, 255	0.9575	0.6395	0.6395	1
	(v)	0, 49, 72, 112, 144, 174, 207, 255	0.9759	0.7051	0.7051	
	(vi)	0, 49, 77, 107, 146, 172, 206, 255	0.9771	0.7054	0.7054	
	(vii)	0, 49, 90, 121, 146, 172, 206, 255	0.9812	0.7747	0.7747	2
	(viii)	0, 49, 81, 102, 146, 172, 206, 255	0.9754	0.7047	0.7047	
	(ix)	0, 49, 90, 120, 146, 172, 206, 255	0.9812	0.7722	0.7722	3
	(x)	0, 49, 82, 120, 146, 172, 206, 255	0.9794	0.7426	0.7425	4

2. *MUSIG-guided segmentation evaluation*: In the same way, the heuristically selected class boundaries with the conventional MUSIG activation function for greyscale images are shown in Table 5.7. The same evaluation functions are applied and the results are reported for the individual set of class boundaries. The corresponding Q values evaluated after the segmentation process are tabulated in Table 5.8. It is quite discernible from Tables 5.4, 5.5, 5.6 and 5.8 that the fitness values derived by the multi-objective genetic algorithm-based Opti-MUSIG activation function for greyscale images are better than those obtained by the conventional MUSIG activation function.

The pareto optimal set of optimised class boundaries and corresponding evaluated segmentation quality measures (ρ, F, F') for eight classes of greyscale Lena image are tabulated in Table 5.1 and the corresponding fixed class boundaries and the evaluated segmentation quality measures for eight classes of the same image are shown in Table 5.7. Most of the cases, it is observed that the segmentation quality measures derived by the narrated method are better than the same deduced by the conventional fixed class responses. It is also detected that one of the segmentation quality measures obtained by the narrated method may not be good than the same found by the conventional MUSIG activation function. The same thing is also observed when we go though the Tables 5.2 and 5.3 for greyscale Baboon and greyscale Brain MRI images, respectively. In these tables, the pareto optimal set of optimised class boundaries and corresponding evaluated segmentation quality measure (ρ, F, F') values are reported. The fixed class responses and the corresponding evaluated segmentation quality measure values are tabulated in Table 5.7 for colour baboon and colour Brain MRI image. The narrated method does better segmentation of the Baboon and Brain MRI images than the MUSIG activation function with respect to the three segmentation quality measures (ρ, F, F'). The evaluated segmentation quality measure Q of each optimised class boundaries are tabulated in Tables 5.4, 5.5 and 5.6 for the test images and the same of the fixed class boundaries are reported in Table 5.8. A detailed scrutiny revealed that the Q reported in Tables 5.4, 5.5 and 5.6 are much

Table 5.2 Pareto optimal set of optimised class boundaries and corresponding evaluated segmentation quality measures for eight classes of greyscale Baboon image

Set no		Class level	ρ	F	F'	κ
1	(i)	0, 56, 91, 115, 139, 162, 185, 255	0.9762	0.7831	0.7831	4
	(ii)	0, 56, 91, 115, 139, 162, 187, 255	0.9767	0.8405	0.8405	
	(iii)	0, 56, 89, 115, 136, 162, 185, 255	0.9753	0.7787	0.7787	
	(iv)	0, 60, 91, 115, 139, 162, 187, 255	0.9774	0.9386	0.9386	3
	(v)	0, 56, 91, 123, 139, 162, 185, 255	0.9735	0.7753	0.7753	1
	(vi)	0, 60, 89, 114, 139, 162, 187, 255	0.9768	0.9333	0.9333	
	(vii)	0, 56, 74, 114, 138, 162, 185, 255	0.9692	0.7498	0.7498	
	(viii)	0, 56, 89, 115, 146, 161, 185, 255	0.9708	0.7751	0.7751	
	(ix)	0, 60, 90, 115, 139, 162, 187, 255	0.9769	0.9356	0.9356	
	(x)	0, 51, 74, 114 146, 162, 185, 255	0.9641	0.7379	0.7379	
	(xi)	0, 56, 89, 115, 139, 162, 185, 255	0.9758	0.7787	0.7787	
	(xii)	0, 56, 90, 115, 139, 162, 185, 255	0.9759	0.7814	0.7814	
	(xiii)	0, 56, 91, 121, 139, 162, 185, 255	0.9747	0.7780	0.7780	2
2	(i)	0, 56, 102, 124, 142, 161, 187, 255	0.9722	0.7416	0.7416	2
	(ii)	0, 56, 94, 117, 138, 161, 187, 255	0.9760	0.7876	0.7876	
	(iii)	0, 61, 94, 118, 137, 159, 187, 255	0.9770	0.9299	0.9299	3
	(iv)	0, 56, 102, 124, 143, 161, 187, 255	0.9721	0.7306	0.7306	1
	(v)	0, 53, 94, 120, 138, 161, 187, 255	0.9744	0.7834	0.7834	4
	(vi)	0, 56, 90, 117, 138, 161, 187, 255	0.9760	0.7852	0.7853	
	(vii)	0, 61, 90, 114, 137, 159, 187, 255	0.9772	0.9318	0.9318	
	(viii)	0, 53, 90, 114, 138, 161, 187, 255	0.9755	0.7842	0.7842	
	(ix)	0, 57, 90, 113, 138, 161, 187, 255	0.9766	0.8727	0.8727	
3	(i)	0, 54, 91, 115, 138, 158, 185, 255	0.9752	0.7883	0.7883	
	(ii)	0, 59, 89, 115, 138, 158, 185, 255	0.9762	0.8818	0.8818	4
	(iii)	0, 57, 89, 114, 138, 158, 185, 255	0.9762	0.8725	0.8725	
	(iv)	0, 55, 89, 115, 138, 159, 185, 255	0.9759	0.8391	0.8391	
	(v)	0, 52, 91, 115, 136, 158, 185, 255	0.9749	0.7853	0.7853	3
	(vi)	0, 52, 91, 115, 138, 158, 186, 255	0.9749	0.7855	0.7855	
	(vii)	0, 55, 91, 123, 142, 158, 185, 255	0.9723	0.7820	0.7820	1
	(viii)	0, 60, 91, 115, 137, 159, 185, 255	0.9772	0.8868	0.8868	2
4	(i)	0, 59, 91, 115, 138, 160, 186, 255	0.9767	0.9247	0.9247	
	(ii)	0, 49, 99, 128, 142, 159, 187, 255	0.9680	0.7697	0.7697	1
	(iii)	0, 59, 92, 115, 138, 160, 186, 255	0.9770	0.9268	0.9268	
	(iv)	0, 59, 92, 115, 136, 160, 186, 255	0.9769	0.9261	0.9261	2
	(v)	0, 63, 92, 115, 138, 160, 185, 255	0.9775	1.0000	1.0000	4
	(vi)	0, 56, 88, 115, 138, 159, 187, 255	0.9761	0.8345	0.8345	
	(vii)	0, 51, 92, 119, 143, 160, 187, 255	0.9727	0.8220	0.8219	
	(viii)	0, 63, 92, 115, 136, 160, 187, 255	0.9772	0.9953	0.9953	3
	(ix)	0, 58, 92, 115, 138, 160, 186, 255	0.9765	0.9221	0.9221	
	(x)	0, 58, 88, 115, 138, 160, 186, 255	0.9762	0.8666	0.8666	

Table 5.3 Pareto optimal set of optimised class boundaries and corresponding evaluated segmentation quality measures for eight classes of greyscale Brain MRI image

Set no		Class level	ρ	F	F'	κ
1	(i)	0, 54, 89, 110, 127, 141, 164, 255	0.9951	0.6106	0.6106	
	(ii)	0, 43, 77, 96, 110, 127, 164, 255	0.9963	0.7318	0.7318	2
	(iii)	0, 55, 96, 110, 131, 148, 164, 255	0.9945	0.5955	0.5955	
	(iv)	0, 48, 76, 96, 110, 128, 166, 255	0.9961	0.7198	0.7198	3
	(v)	0, 50, 89, 110, 127, 141, 164, 255	0.9953	0.6112	0.6112	
	(vi)	0, 46, 81, 95, 113, 128, 166, 255	0.9961	0.7182	0.7182	1
	(vii)	0, 58, 96, 110, 127, 141, 164, 255	0.9943	0.5644	0.5644	
	(viii)	0, 50, 78, 96, 110, 127, 164, 255	0.9960	0.6720	0.6720	4
2	(i)	0, 47, 80, 98, 113, 128, 165, 255	0.9961	0.7130	0.7130	2
	(ii)	0, 55, 84, 99, 113, 128, 165, 255	0.9955	0.6407	0.6407	4
	(iii)	0, 60, 95, 113, 128, 145, 165, 255	0.9944	0.5645	0.5645	
	(iv)	0, 55, 95, 112, 129, 145, 164, 255	0.9948	0.5958	0.5958	
	(v)	0, 53, 84, 99, 113, 128, 165, 255	0.9957	0.6507	0.6507	3
	(vi)	0, 55, 97, 112, 129, 143, 166, 255	0.9947	0.5951	0.5951	
	(vii)	0, 60, 95, 113, 128, 145, 164, 255	0.9944	0.5645	0.5645	
	(viii)	0, 55, 97, 112, 129, 143, 166, 255	0.9947	0.5951	0.5951	
	(ix)	0, 52, 93, 113, 128, 145, 164, 255	0.9951	0.6040	0.6040	
	(x)	0, 54, 95, 112, 129, 145, 164, 255	0.9949	0.5958	0.5958	
	(xi)	0, 78, 97, 112, 128, 143, 166, 255	0.9902	0.4772	0.4772	1

Table 5.4 The evaluated segmentation quality measure Q of the each optimised class boundaries of Table 5.1

Set	No	(i)	(ii)	(iii)	(iv)	(v)	(vi)	(vii)
1	Q	0.8659	0.8716	0.7833	0.8596	0.7812	0.7831	0.7950
	No	(viii)	(ix)	(x)	(xi)			
	Q	0.8678	0.7949	0.8002	0.8490			
2	No	(i)	(ii)	(iii)	(iv)	(v)	(vi)	(vii)
	Q	0.8792	0.9741	0.8761	0.8660	0.8883	0.8822	0.9491
	No	(viii)	(ix)	(x)	(xi)	(xii)	(xiii)	
	Q	0.8723	0.8774	0.8822	0.7747	0.8678	0.9146	
3	No	(i)	(ii)	(iii)	(iv)	(v)	(vi)	(vii)
	Q	0.9347	0.8500	0.8671	0.9199	0.7807	0.9277	0.7754
	No	(viii)	(ix)	(x)	(xi)			
	Q	0.9591	0.8718	0.9507	1.0000			
4	No	(i)	(ii)	(iii)	(iv)	(v)	(vi)	(vii)
	Q	0.9451	0.8895	0.9262	0.7576	0.9434	0.9450	0.8549
	No	(viii)	(ix)	(x)				
	Q	0.9585	0.8562	0.8710				

Table 5.5 The evaluated segmentation quality measure Q of the each optimised class boundaries of Table 5.2

Set	No	(i)	(ii)	(iii)	(iv)	(v)	(vi)	(vii)
1	Q	0.7938	0.7955	0.7999	0.7904	0.7584	0.8059	0.8729
	No	(viii)	(ix)	(x)	(xi)	(xii)	(xiii)	
	Q	0.8275	0.7980	0.9130	0.8032	0.8009	0.7655	
2	No	(i)	(ii)	(iii)	(iv)	(v)	(vi)	(vii)
	Q	0.7394	0.7768	0.7515	0.7348	0.7704	0.7888	0.7819
	No	(viii)	(ix)					
	Q	0.8109	0.8060					
3	No	(i)	(ii)	(iii)	(iv)	(v)	(vi)	(vii)
	Q	0.7827	0.7816	0.7842	0.7866	0.7766	0.7828	0.7550
	No	(viii)						
	Q	0.7724						
4	No	(i)	(ii)	(iii)	(iv)	(v)	(vi)	(vii)
	Q	0.7859	0.7452	0.7797	0.7721	0.7728	0.7914	0.7847
	No	(viii)	(ix)	(x)				
	Q	0.7726	0.7827	0.7947				

Table 5.6 The evaluated segmentation quality measure Q of the each optimised class boundaries of Table 5.3

Set	No	(i)	(ii)	(iii)	(iv)	(v)	(vi)	(vii)
1	Q	0.9744	0.9572	0.9765	0.9574	0.9744	0.9569	0.9821
	No	(viii)						
	Q	0.9686						
2	No	(i)	(ii)	(iii)	(iv)	(v)	(vi)	(vii)
	Q	0.9547	0.9660	0.9825	0.9756	0.9658	0.9732	0.9825
	No	(viii)	(ix)	(x)	(xi)			
	Q	0.9732	0.9744	0.9756	0.9534			

better that the same tabulated in Table 5.8. So, it can be concluded that the MOGA-based OptiMUSIG activation function can do better greyscale segmentation than the conventional MUSIC activation function.

5.2.3   Image Segmentation Outputs

The segmented greyscale and true colour output images obtained for the $K = 8$ classes, with the narrated optimised approach vis-a-vis those obtained with the heuristically chosen class boundaries, are presented in this section.

Table 5.7   Fixed class boundaries and corresponding evaluated segmentation quality measures for eight classes of greyscale Baboon and Peppers image

Image	Set no	Class level	ρ	F	F'
Lena	1	0, 50, 80, 105, 130, 165, 205, 255	0.9790	0.7845	0.7845
	2	0, 60, 75, 90, 125, 155, 200, 255	0.9705	0.9335	0.9335
	3	0, 40, 60, 85, 120, 160, 180, 255	0.9664	1.0000	1.0000
	4	0, 45, 65, 80, 125, 155, 185, 255	0.9684	0.9956	0.99
Baboon	1	0, 50, 90, 110, 140, 165, 190, 255	0.9709	0.9040	0.9040
	2	0, 60, 95, 120, 142, 160, 200, 255	0.9682	0.8437	0.8437
	3	0, 40, 80, 110, 150, 180, 210, 255	0.9503	0.6476	0.6476
	4	0, 35, 70, 115, 135, 155, 195, 255	0.9576	0.9031	0.9985
MRI	1	0, 25, 50, 75, 85, 110, 170, 255	0.9914	1.0000	1.0000
	2	0, 30, 60, 75, 90, 105, 160, 255	0.9875	0.9848	0.9848
	3	0, 28, 58, 78, 88, 108, 165, 255	0.9902	0.9408	0.9408
	4	0, 33, 63, 83, 93, 113, 155, 255	0.9928	0.9825	0.9825

Table 5.8   The evaluated segmentation quality measure Q of the each fixed class boundaries of Table 5.7

Image	Set no	Q	Image	Set no	Q	Image	Set no	Q
Lena	1	0.8675	Baboon	1	0.8670	MRI	1	0.9991
	2	0.9102		2	0.7310		2	0.9879
	3	0.9073		3	1.0000		3	1.0000
	4	0.9004		4	0.8760		4	0.9814

5.2.3.1   Multi-objective Genetic Algorithm-Based OptiMUSIG Guided Segmented Outputs

The first four better results of each Pareto optimal set are applied in the networks. The segmented multilevel greyscale test images obtained with the MLSONN architecture using the multi-objective genetic algorithm-based OptiMUSIG activation function for $K = 8$ are shown in Figs. 5.2, 5.3 and 5.4.

5.2.3.2   MUSIG Guided Segmented Outputs

The segmented multilevel greyscale test images obtained with the MLSONN architecture characterised by the conventional MUSIG activation employing fixed class responses for $K = 8$, are shown in Figs. 5.5 and 5.6.

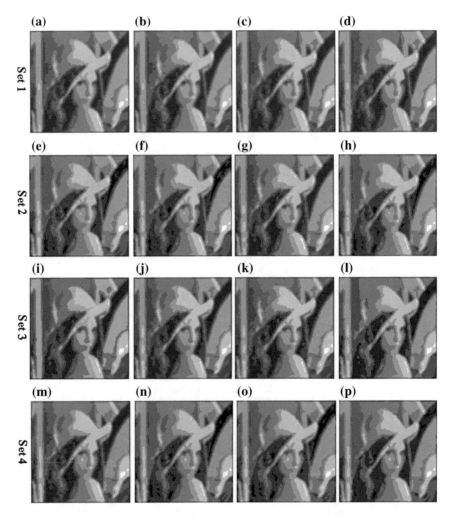

Fig. 5.2 8-class segmented 128×128 greyscale Lena image with the optimised class levels referring to (**a–d**) set 1 (**e–h**) set 2 (i-l) set 3 (m-p) set 4 of Table 5.1 for first four better quality measure Q with multi-objective genetic algorithm-based OptiMUSIG activation function

From the results obtained, it is evident that the multi-objective genetic algorithm-based OptiMUSIG activation function for greyscale images outperforms its conventional MUSIG counterpart as regards to the segmentation quality of the images for different number of classes. Moreover, since the narrated approach incorporates the image heterogeneity, it can handle a wide variety of image intensity distribution prevalent in the real life.

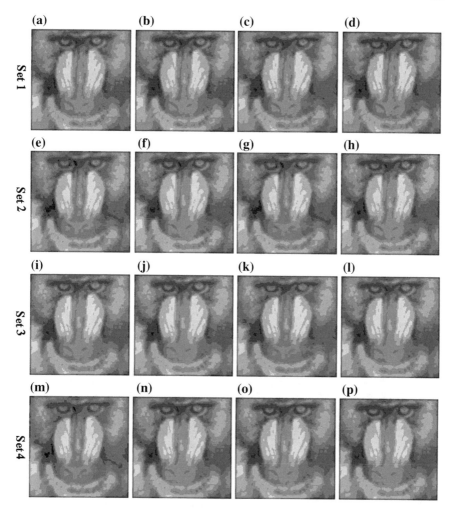

Fig. 5.3 8-class segmented 128×128 greyscale Baboon image with the optimised class levels referring to (**a–d**) set 1 (**e–h**) set 2 (i-l) set 3 (m-p) set 4 of Table 5.2 for first four better quality measure Q with multi-objective genetic algorithm-based OptiMUSIG activation function

5.3 NSGA-II-Based OptiMUSIG Activation Function

In this section of this chapter, a popular and efficient multi-objective-based algorithm, NSGA-II, is applied to generate the optimised class levels. The detailed discussion of the OptiMUSIG activation function is illustrated in the Chap. 3. As it is already observed that the single objective-based OptiMUSIG activation function is not very much effective for different objective functions to segment the multilevel greyscale images. To overcome this problem, a NSGA-II-based OptiMUSIG activation function $(\bar{f}_{Opti_{NSGA}}(\bar{x}))$ is presented in this chapter and it is denoted as [100, 242, 287]

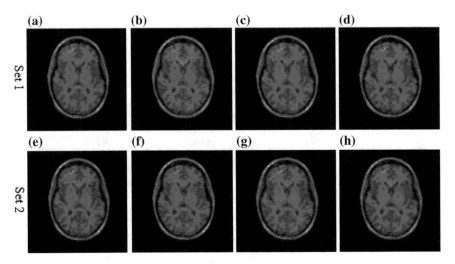

Fig. 5.4 8-class segmented 128×128 greyscale Brain MRI image with the optimised class levels referring to (**a–d**) set 1 (**e–h**) set 2 of Table 5.3 for first four better quality measure Q with multi-objective genetic algorithm-based OptiMUSIG activation function

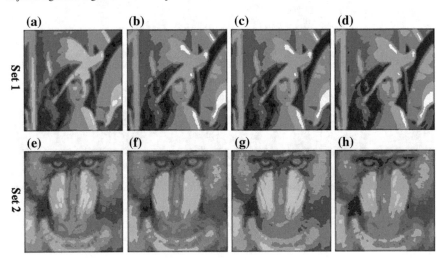

Fig. 5.5 8-class segmented 128×128 (**a–d**) greyscale Lena image and (**e–h**) greyscale Baboon image with the fixed class levels of Table 5.7 with the MUSIG activation function

$$\bar{f}_{Opti_{NSGA}}(\overline{x}) = [\bar{f}_{Opti_{F_1}}(\overline{x}), \bar{f}_{Opti_{F_2}}(\overline{x}), \ldots, \bar{f}_{Opti_{F_n}}(\overline{x})]^T \qquad (5.1)$$

subject to $g_{Opti_{F_i}}(\overline{x}) \geq 0, i = 1, 2, \ldots, m$ and $h_{Opti_{F_i}}(\overline{x}) = 0, i = 1, 2, \ldots, p$ where, $f_{Opti_{F_i}}(\overline{x}) \geq 0, i = 1, 2, \ldots, n$ are different objective functions in the NSGA-II-based

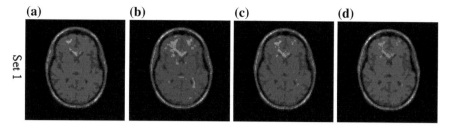

Fig. 5.6 8-class segmented 128 × 128 (**a–d**) greyscale Brain MRI image with the fixed class levels of Table 5.7 with the MUSIG activation function

OptiMUSIG activation function. The inequality and equality constraints of this function are denoted as $g_{Opti_{F_i}}(\overline{x}) \geq 0, i = 1, 2, \ldots, m$ and $h_{Opti_{F_i}}(\overline{x}) = 0, i = 1, 2, \ldots, p$, respectively [100, 242, 287].

5.3.1 Multilevel Greyscale Image Segmentation by NSGA-II-Based OptiMUSIG Activation Function

In this section, the narrated methodology of the multilevel greyscale image segmentation process has been discussed using the NSGA-II-based OptiMUSIG activation function [100, 242, 287] with a MLSONN architecture and it has been operated in three phases. The different phases of this method are depicted in Fig. 5.7 and elaborated in the following subsections.

Fig. 5.7 Flow diagram of greyscale image segmentation using NSGA-II-based OptiMUSIG activation function

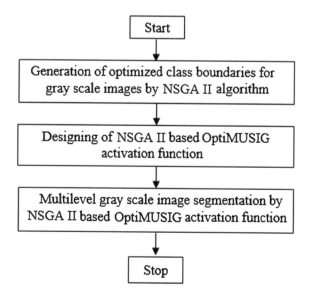

5.3.1.1 Generation of Optimised Class Boundaries for Greyscale Images by NSGA II Algorithm

This is the most important phase of this approach. The NSGA II-based optimisation procedure is employed to generate the optimised class boundaries $(c_{\gamma_{opt}})$ of the NSGA-II-based OptiMUSIG activation function. The detailed description of this stage is described as follows:

- *Initialization phase*: The pixel intensity levels of the greyscale image and the number of classes (K) to be segmented are supplied as inputs to the NSGA-II-based optimisation procedure in this phase.
- *Chromosome representation and population generation*: The real numbered chromosomes are applied in the NSGA-II algorithm and the real numbers are generated randomly for developing the optimised class boundaries from the input image information content. A population size of 200 has been utilised in this treatment.
- *Fitness computation*: In this phase, three segmentation evaluation metrices (ρ, F, F') given in Eqs. 3.25, 3.28 and 3.29 respectively, are applied as the fitness functions. These functions are applied to evaluate the quality of the segmented images in this NSGA-II-based optimisation procedure.
- *Genetic operators*: Selection, crossover and mutation, three common genetic operators, are applied in this method. The selection operator, the crowded binary tournament selection, is applied in NSGA-II. After selection, the selected chromosomes are put in the mating pool for the crossover and mutation operation. The crossover probability is taken as 0.8 and the mutation probability is chosen as 0.1 in this approach. In the next generation of NSGA II, the non-dominated solutions among the parent and child populations are propagated. The details of the different genetic processes are depicted in [98]. In this approach, this process continued for 10^4 iterations. After the last generation, the near-pareto optimal strings render the desired solutions.
- *Selecting a solution from the non-dominated set*: A particular solution has to be selected from the set of non-dominated solutions having rank one in the NSGA-II algorithm to design the OptiMUSIG activation function. For that reason, the empirical measure (EF), Q index [256] has been employed to select the better chromosomes from the selected chromosome pool in this approach.

5.3.1.2 Designing of NSGA-II-Based OptiMUSIG Activation Function

The NSGA-II-based OptiMUSIG activation function has been designed with the help of the optimised class boundaries $(c_{\gamma_{opt}})$ selected in the Pareto optimal set. The $\alpha_{\gamma_{opt}}$ parameters are determined using the optimised $(c_{\gamma_{opt}})$ by the Eq. 3.23. These parameters are further employed to obtain the different transition levels of the NSGA-II-based OptiMUSIG activation function.

5.3.1.3 Multilevel Greyscale Image Segmentation by NSGA-II-Based OptiMUSIG Activation Function

The real-life multilevel greyscale images are segmented by the single MLSONN architecture characterised by NSGA-II-based OptiMUSIG activation function in the last phase of this approach. The succeeding network layers propagates the processed input information. Different grey level responses of the input image have been assessed by the neurons of the different layers of the MLSONN architecture. As it is a self-supervised neural network, the MLSONN architecture has no *a priori* knowledge about the outputs. The subnormal linear index of fuzziness [72] is employed to determine the system errors at the output layer of the MLSONN architecture. These errors are used to adjust the interconnection weights between the different layers using the standard backpropagation algorithm [37, 47]. The outputs of the output layer of the network are transferred to the input layer of the network to minimise the system errors for further processing. The original input image gets segmented into different multilevel regions depending upon the optimised transition levels of the OptiMUSIG activation function after the self-supervision of the network attains stabilisation.

5.3.2 Result Analysis

Two real-life images, viz. Baboon and Peppers and a Brain Neuroanatomy (MRI) image [253], each of dimensions 128×128, have been used to demonstrate the multilevel greyscale image segmentation approach using the MLSONN architecture guided by NSGA-II-based OptiMUSIG activation function. Experiments have been carried with $K = \{6, 8\}$ classes. The activation function has been designed with a fixed slope, $\lambda = \{2, 4\}$. But the results are reported for eight classes and a fixed slope of $\lambda = 4$. The segmentation efficiency of the narrated approach is demonstrated in the following subsection and it is compared with the segmentation derived by means of the conventional MUSIG activation function with same number of class responses and heuristic class levels. In the next section, the quantitative comparison of the above- mentioned methods is manifested and after that, the comparative study as regards to its efficacy in the segmentation of test images of the segmented outputs by the present methods and the conventional MUSIG activation function-based methods is detailed.

5.3.2.1 Quantitative Performance Analysis of Segmentation

In this section, the efficiency of the NSGA-II-based OptiMUSIG and the conventional MUSIG activation functions for $K = 8$ has been exemplified quantitatively. Three segmentation efficiency functions, viz. correlation coefficient (ρ), and two empirical evaluation function (F and F') have been employed in the NSGA-II-based approach to generate the activation functions and another empirical evaluation function (Q) has been employed for the selection of the chromosome that is applied in the corresponding network in the present method. The results obtained with the NSGA-II-based OptiMUSIG have been discussed in the next subsection and the corresponding results obtained with the conventional fixed class response based MUSIG activation function are furnished in the next to next subsection.

1. *Segmentation evaluation using NSGA-II-based OptiMUSIG activation function for greyscale images*: In Tables 5.9, 5.10 and 5.11, the optimised class boundaries ($c_{\gamma_{opt}}$) for the greyscale images obtained using NSGA-II-based optimisation with the three evaluation functions (ρ, F and F') are tabulated. Three fitness values (ρ, F and F') of individual chromosome set are also reported in these tables. Two sets of Pareto optimal chromosomes of all the images are tabulated in the above mentioned tables. The Q values of individual set of class boundaries for the test images which have been used for the gradation system for that particular set of class boundaries are presented in Tables 5.12, 5.13 and 5.14. The last columns of the Tables 5.9, 5.10 and 5.11 show the quality measures κ (graded on a scale of 1 (best) onwards) obtained by the segmentation of the test images based on the corresponding set of optimised class boundaries. The first four good Q valued chromosomes are indicated in those tables for easy reckoning. In this section, all evaluation function values are tabulated in normalised form.

2. *Segmentation evaluation using MUSIG activation function for greyscale images and colour images*: The heuristically selected class boundaries are applied in the conventional MUSIG activation function for greyscale images and they are depicted in Table 5.15. The evaluation procedure is exercised for the individual set of class boundaries along with the same evaluation functions and the results are reported in those tables. After the segmentation process using the MUSIG activation function, the quality of the segmented images are evaluated using the Q evaluation function and the results are tabulated in Table 5.16 for greyscale.

It is quite evident from Tables 5.12, 5.13, 5.14 and 5.16 that the fitness values derived by the NSGA II based OptiMUSIG activation function for gray scale images are better than those obtained by the conventional MUSIG activation function. In most of the cases, the Q value in Tables 5.12, 5.13 and 5.14 are lower than the same value in Table 5.16 for all images. The lower value of Q denotes the better segmentation as it is known from the previous discussion. It is discernible from those tables that the image segmentation done by the presented method generates the lower Q value than the image segmentation done by the heuristically selected class levels. This can also be proved from Tables 5.9, 5.10, 5.11 and 5.15 if we observe these tables minutely. In those tables, the ρ, F and F' values derived by the NSGA II based OptiMUSIG activation function for greyscale images are better than those derived by the conventional MUSIG activation function for the respective images.

Table 5.9 NSGA-II-based set of optimised class boundaries and corresponding evaluated segmentation quality measures for eight classes of greyscale Baboon image

Set no		Class level	ρ	F	F'	κ
1	(i)	4, 70, 95, 116, 137, 159, 184, 217	0.9764	0.4904	0.1734	
	(ii)	4, 78, 89, 114, 127, 147, 154, 217	0.9442	0.4447	0.1572	1
	(iii)	4, 68, 95, 116, 137, 159, 182, 217	0.9760	0.4880	0.1725	
	(iv)	4, 74, 93, 118, 132, 159, 163, 217	0.9543	0.4462	0.1578	2
	(v)	4, 76, 93, 118, 135, 159, 174, 217	0.9576	0.4531	0.1602	
	(vi)	4, 74, 95, 116, 136, 159, 167, 217	0.9570	0.4514	0.1596	
	(vii)	4, 74, 93, 118, 135, 159, 172, 217	0.9666	0.4692	0.1659	
	(viii)	4, 72, 95, 116, 137, 159, 182, 217	0.9753	0.4850	0.1715	
	(ix)	4, 72, 95, 118, 137, 161, 180, 217	0.9677	0.4705	0.1663	
	(x)	4, 71, 95, 116, 136, 159, 167, 217	0.9759	0.4863	0.1719	
	(xi)	4, 70, 95, 114, 136, 157, 167, 217	0.9740	0.4820	0.1704	4
	(xii)	4, 70, 95, 116, 137, 161, 180, 217	0.9548	0.4475	0.1582	
	(xiii)	4, 72, 95, 116, 137, 159, 176, 217	0.9560	0.4493	0.1589	
	(xiv)	4, 72, 95, 116, 137, 161, 176, 217	0.9627	0.4610	0.1630	
	(xv)	4, 70, 95, 116, 137, 159, 180, 217	0.9635	0.4638	0.1640	
	(xvi)	4, 70, 95, 116, 138, 159, 182, 217	0.9746	0.4832	0.1708	
	(xvii)	4, 74, 95, 116, 136, 161, 176, 217	0.9717	0.4763	0.1684	
	(xviii)	4, 72, 93, 116, 134, 157, 167, 217	0.9718	0.4773	0.1688	
	(xix)	4, 71, 95, 116, 136, 155, 167, 217	0.9638	0.4646	0.1643	3
2	(i)	4, 78, 84, 115, 121, 148, 152, 217	0.9358	0.4327	0.1530	2
	(ii)	4, 68, 95, 117, 138, 160, 185, 217	0.9767	0.4949	0.1750	
	(iii)	4, 68, 93, 115, 136, 153, 165, 217	0.9626	0.4672	0.1652	
	(iv)	4, 78, 95, 117, 135, 160, 177, 217	0.9682	0.4731	0.1672	
	(v)	4, 71, 95, 115, 134, 155, 165, 217	0.9619	0.4643	0.1641	
	(vi)	4, 71, 95, 115, 134, 153, 165, 217	0.9756	0.4865	0.1720	
	(vii)	4, 68, 96, 117, 137, 160, 181, 217	0.9614	0.4611	0.1630	
	(viii)	4, 71, 93, 115, 134, 153, 163, 217	0.9748	0.4830	0.1708	
	(ix)	4, 72, 95, 117, 137, 160, 181, 217	0.9738	0.4808	0.1700	
	(x)	4, 68, 96, 117, 138, 158, 183, 217	0.9713	0.4749	0.1679	
	(xi)	4, 74, 95, 117, 137, 160, 177, 217	0.9375	0.4346	0.1537	
	(xii)	4, 69, 95, 115, 134, 155, 163, 217	0.9719	0.4766	0.1685	4
	(xiii)	4, 78, 86, 115, 121, 148, 154, 217	0.9595	0.4571	0.1616	3
	(xiv)	4, 78, 88, 115, 121, 148, 152, 217	0.9578	0.4534	0.1603	1
	(xv)	4, 72, 95, 117, 137, 160, 179, 217	0.9596	0.4596	0.1625	
	(xvi)	4, 73, 95, 117, 137, 160, 177, 217	0.9761	0.4887	0.1728	
	(xvii)	4, 72, 95, 117, 134, 157, 163, 217	0.9481	0.4486	0.1586	

Table 5.10 NSGA-II-based set of optimised class boundaries and corresponding evaluated segmentation quality measures for eight classes of greyscale Peppers image

Set no		Class level	ρ	F	F'	κ
1	(i)	0, 40, 78, 107, 139, 168, 195, 248	0.9831	0.7697	0.2721	
	(ii)	0, 52, 71, 111, 140, 166, 189, 248	0.9780	0.7317	0.2587	
	(iii)	0, 42, 78, 107, 139, 168, 195, 248	0.9831	0.7653	0.2706	
	(iv)	0, 42, 75, 107, 139, 168, 195, 248	0.9830	0.7600	0.2687	
	(v)	0, 44, 78, 109, 140, 168, 195, 248	0.9830	0.7633	0.2699	2
	(vi)	0, 52, 71, 111, 140, 166, 191, 248	0.9788	0.7327	0.2591	
	(vii)	0, 54, 73, 111, 139, 166, 191, 248	0.9789	0.7345	0.2597	4
	(viii)	0, 44, 78, 109, 140, 166, 195, 248	0.9808	0.7400	0.2616	1
	(ix)	0, 47, 76, 109, 141, 166, 193, 248	0.9805	0.7385	0.2611	
	(x)	0, 47, 73, 109, 140, 166, 191, 248	0.9816	0.7445	0.2632	
	(xi)	0, 49, 73, 109, 139, 166, 191, 248	0.9829	0.7568	0.2676	
	(xii)	0, 44, 76, 109, 139, 166, 195, 248	0.9823	0.7495	0.2650	
	(xiii)	0, 46, 76, 109, 140, 166, 191, 248	0.9820	0.7466	0.2640	3
	(xiv)	0, 44, 76, 109, 140, 168, 195, 248	0.9824	0.7505	0.2654	
	(xv)	0, 46, 75, 109, 139, 166, 193, 248	0.9825	0.7515	0.2657	
	(xvi)	0, 49, 73, 109, 141, 166, 193, 248	0.9827	0.7536	0.2665	
	(xvii)	0, 44, 75, 107, 139, 168, 195, 248	0.9826	0.7528	0.2661	
	(xviii)	0, 46, 75, 109, 141, 166, 195, 248	0.9826	0.7520	0.2659	
	(xix)	0, 47, 75, 109, 140, 166, 193, 248	0.9796	0.7357	0.2601	
	(xx)	0, 47, 76, 109, 140, 168, 195, 248	0.9828	0.7565	0.2675	
2	(i)	0, 42, 76, 110, 138, 168, 196, 248	0.9700	0.7653	0.2706	
	(ii)	0, 61, 67, 114, 137, 170, 187, 248	0.9829	0.7188	0.2541	
	(iii)	0, 42, 76, 108, 142, 166, 196, 248	0.9828	0.7610	0.2691	
	(iv)	0, 42, 76, 108, 140, 168, 194, 248	0.9822	0.7587	0.2683	
	(v)	0, 50, 76, 108, 140, 166, 194, 248	0.9819	0.7510	0.2655	4
	(vi)	0, 44, 76, 110, 140, 166, 193, 248	0.9811	0.7484	0.2646	
	(vii)	0, 50, 76, 108, 140, 166, 191, 248	0.9825	0.7441	0.2631	1
	(viii)	0, 52, 71, 110, 140, 166, 191, 248	0.9813	0.7546	0.2668	
	(ix)	0, 50, 73, 110, 140, 166, 193, 248	0.9828	0.7457	0.2637	
	(x)	0, 50, 76, 110, 140, 166, 193, 248	0.9825	0.7567	0.2675	2
	(xi)	0, 44, 78, 108, 142, 166, 194, 248	0.9799	0.7537	0.2665	
	(xii)	0, 50, 76, 108, 140, 166, 193, 248	0.9784	0.7377	0.2608	3
	(xiii)	0, 52, 71, 110, 140, 166, 193, 248	0.9826	0.7318	0.2587	
	(xiv)	0, 44, 76, 108, 142, 166, 196, 248	0.9797	0.7564	0.2674	
	(xv)	0, 44, 76, 110, 140, 168, 194, 248	0.9713	0.7357	0.2601	
	(xvi)	0, 52, 71, 112, 139, 168, 191, 248	0.9792	0.7214	0.2551	
	(xvii)	0, 42, 76, 108, 142, 166, 194, 248	0.9700	0.7335	0.2593	

Table 5.11 NSGA-II-based set of optimised class boundaries and corresponding evaluated segmentation quality measures for eight classes of greyscale Brain MRI image

Set no		Class level	ρ	F	F'	κ
1	(i)	0, 41, 65, 91, 108, 131, 142, 255	0.9952	0.9186	0.9186	
	(ii)	0, 41, 62, 89, 110, 127, 158, 255	0.9962	0.8864	0.8864	
	(iii)	0, 28, 51, 74, 93, 111, 129, 255	0.9925	0.8582	0.8582	4
	(iv)	0, 41, 62, 89, 108, 129, 164, 255	0.9964	0.8338	0.8338	
	(v)	0, 33, 51, 74, 93, 111, 129, 255	0.9924	0.8516	0.8516	3
	(vi)	0, 34, 53, 74, 93, 109, 129, 255	0.9923	0.8133	0.8133	
	(vii)	0, 41, 62, 89, 108, 131, 142, 255	0.9952	0.9408	0.9408	
	(viii)	0, 33, 51, 74, 93, 110, 134, 255	0.9940	0.8540	0.8540	1
	(ix)	0, 28, 51, 74, 93, 112, 136, 255	0.9942	0.8603	0.8603	2
2	(i)	0, 34, 61, 89, 110, 128, 156, 255	0.9965	0.9018	0.9018	
	(ii)	0, 33, 53, 75, 100, 113, 133, 255	0.9943	0.8337	0.8337	2
	(iii)	0, 49, 71, 94, 111, 128, 156, 255	0.9960	0.7983	0.7983	
	(iv)	0, 42, 61, 89, 110, 128, 156, 255	0.9962	0.8444	0.8444	
	(v)	0, 49, 55, 83, 100, 111, 131, 255	0.9911	0.7218	0.7218	3
	(vi)	0, 49, 61, 93, 109, 130, 154, 255	0.9952	0.8395	0.8395	
	(vii)	0, 33, 55, 75, 100, 113, 133, 255	0.9943	0.8216	0.8216	1
	(viii)	0, 42, 61, 81, 100, 113, 133, 255	0.9942	0.7229	0.7229	4
	(ix)	0, 42, 65, 93, 111, 128, 156, 255	0.9964	0.8184	0.8184	
	(x)	0, 49, 63, 94, 111, 128, 156 255	0.9957	0.8173	0.8173	

Table 5.12 The evaluated segmentation quality measure Q of the each optimised class boundaries of Table 5.9

Set	No	(i)	(ii)	(iii)	(iv)	(v)	(vi)	(vii)
1	Q	0.3058	0.2494	0.3022	0.2745	0.2872	0.2840	0.2856
	No	(viii)	(ix)	(x)	(xi)	(xii)	(xiii)	(xiv)
	Q	0.3004	0.2975	0.2850	0.2823	0.3048	0.2943	0.2982
	No	(xv)	(xvi)	(xvii)	(xviii)	(xix)		
	Q	0.3012	0.2994	0.2940	0.2824	0.2758		
2	No	(i)	(ii)	(iii)	(iv)	(v)	(vi)	(vii)
	Q	0.2557	0.3044	0.2704	0.2940	0.2709	0.2708	0.2995
	No	(viii)	(ix)	(x)	(xi)	(xii)	(xiii)	(xiv)
	Q	0.2715	0.2977	0.2985	0.2931	0.2701	0.2579	0.2525
	No	(xv)	(xvi)	(xvii)				
	Q	0.2968	0.2933	0.2717				

Table 5.13 The evaluated segmentation quality measure Q of the each optimised class boundaries of Table 5.10

Set	No	(i)	(ii)	(iii)	(iv)	(v)	(vi)	(vii)
1	Q	0.8813	0.8911	0.8800	0.8881	0.8696	0.8943	0.8794
	No	(viii)	(ix)	(x)	(xi)	(xii)	(xiii)	(xiv)
	Q	0.8688	0.8837	0.8828	0.8843	0.8879	0.8762	0.8866
	No	(xv)	(xvi)	(xvii)	(xviii)	(xix)	(xx)	
	Q	0.8851	0.8933	0.8881	0.8918	0.8824	0.8860	
2	No	(i)	(ii)	(iii)	(iv)	(v)	(vi)	(vii)
	Q	0.8841	0.8876	0.8945	0.8850	0.8793	0.8811	0.8718
	No	(viii)	(ix)	(x)	(xi)	(xii)	(xiii)	(xiv)
	Q	0.8931	0.8852	0.8760	0.8814	0.8769	0.8983	0.8943
	No	(xv)	(xvi)	(xvii)				
	Q	0.8846	0.8875	0.8934				

Table 5.14 The evaluated segmentation quality measure Q of the each optimised class boundaries of Table 5.11

Set	No	(i)	(ii)	(iii)	(iv)	(v)	(vi)	(vii)
1	Q	0.9284	0.9341	0.9210	0.9357	0.9209	0.9297	0.9295
	No	(viii)	(ix)					
	Q	0.9193	0.9194					
2	No	(i)	(ii)	(iii)	(iv)	(v)	(vi)	(vii)
	Q	0.9341	0.9152	0.9408	0.9455	0.9310	0.9425	0.9151
	No	(viii)	(ix)	(x)				
	Q	0.9318	0.9425	0.9425				

Table 5.15 Fixed class boundaries and corresponding evaluated segmentation quality measures for eight classes of greyscale Baboon and Peppers image

Image	Set no	Class level	ρ	F	F'
Baboon	1	4, 50, 80, 105, 130, 165, 205, 217	0.9691	0.8849	0.8849
	2	4, 49, 74, 100, 135, 160, 200, 217	0.9724	0.8180	0.8180
	3	4, 36, 76, 105, 142, 172, 212, 217	0.9664	0.9711	0.9711
	4	4, 30, 80, 105, 145, 180, 205, 217	0.9656	1.0000	1.0000
Peppers	1	0, 35, 80, 110, 152, 175, 200, 248	0.9852	0.9187	0.9187
	2	0, 30, 75, 115, 150, 185, 205, 248	0.9825	0.9996	0.9996
	3	0, 40, 78, 112, 155, 188, 208, 248	0.9835	1.0000	1.0000
	4	0, 30, 73, 106, 145, 178, 200, 248	0.9849	0.9445	0.9445
MRI	1	0, 28, 55, 74, 84, 111, 171, 255	0.9918	0.9604	0.9604
	2	0, 30, 60, 75, 88, 105, 162, 255	0.9881	0.9845	0.9845
	3	0, 27, 58, 78, 92, 108, 165, 255	0.9902	0.9295	0.9295
	4	0, 32, 62, 82, 85, 114, 155, 255	0.9920	1.0000	1.0000

Table 5.16 The evaluated segmentation quality measure Q of the each fixed class boundaries of Table 5.15

Image	Set no	Q	Image	Set no	Q	Image	Set no	Q
Baboon	1	0.3590	Peppers	1	0.9142	MRI	1	0.9928
	2	0.3787		2	0.9674		2	1.0000
	3	1.0000		3	1.0000		3	0.9799
	4	0.8502		4	0.9399		4	0.9803

5.3.3   Image Segmentation Outputs

The segmented greyscale output images obtained for the $K = 8$ classes, with the presented optimised approach vis-a-vis those obtained with the heuristically chosen class boundaries, are demonstrated in this section.

5.3.3.1   Segmented Outputs Using Optimised Approach

The networks are applied to segment the test images using the first four better results on the basis of the evaluation metric, Q, value. The segmented multilevel greyscale test images obtained with the MLSONN architecture using the NSGA-II-based OptiMUSIG activation function for $K = 8$ are presented in Figs. 5.8, 5.9 and 5.10.

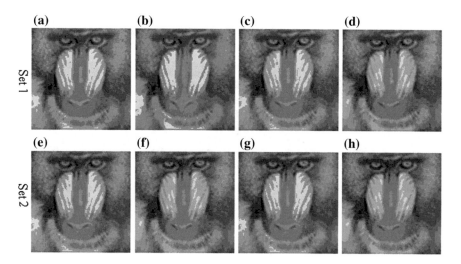

Fig. 5.8 8-class segmented 128×128 greyscale Baboon image with the optimised class levels referring to (**a–d**) set 1 (**e–h**) set 2 of Table 5.9 for first four better quality measure Q with NSGA-II-based OptiMUSIG activation function

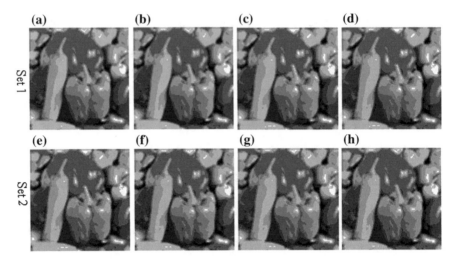

Fig. 5.9 8-class segmented 128×128 greyscale Peppers image with the optimised class levels referring to (**a–d**) set 1 (**e–h**) set 2 of Table 5.10 for first four better quality measure Q with NSGA-II-based OptiMUSIG activation function

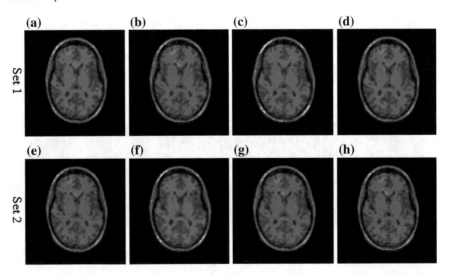

Fig. 5.10 8-class segmented 128×128 greyscale Brain MRI image with the optimised class levels referring to (**a–d**) set 1 (**e–h**) set 2 of Table 5.11 for first four better quality measure Q with NSGA-II-based OptiMUSIG activation function

5.3.3.2 Segmented Outputs Using MUSIG Activation Function

The fixed class responses are employed to generate the segmented multilevel greyscale test images in connection with the MLSONN architecture characterised by the conventional MUSIG activation. The segmented output gray scale images for $K = 8$ are shown in Figs. 5.11 and 5.12.

From the results obtained, the image segmentation method using the conventional MUSIG activation function is outperformed by the image segmentation method by the NSGA-II-based OptiMUSIG activation function for gray scale images as regards to the segmentation quality of the images for the different number of classes. It is also evident that the narrated approach incorporates the image heterogeneity as it can handle a wide variety of image intensity distribution prevalent in real life.

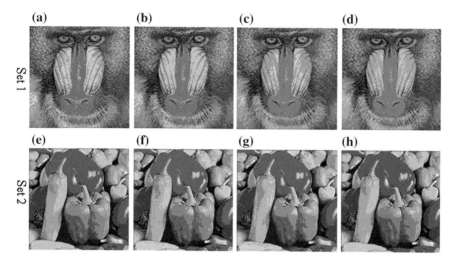

Fig. 5.11 8-class segmented 128×128 (**a–d**) greyscale Baboon image and (**e–h**) greyscale Peppers image with the fixed class levels of Table 5.16 with MUSIG activation function

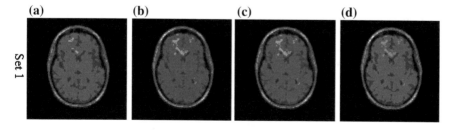

Fig. 5.12 8-class segmented 128×128 (**a–d**) grey scale Brain MRI image with the fixed class levels of Table 5.16 with MUSIG activation function

5.4 Discussions and Conclusion

In this chapter, grey scale image segmentation by the multilayer self organizing neural network (MLSONN) architecture is discussed. The induction of multiscaling capabilities in the MLSONN network is resorted by the optimised multilevel MUSIG activation function. The generation of the OptiMUSIG activation function based on the multi-objective genetic algorithm is reported in the first part of this chapter and a NSGA II based OptiMUSIG activation function is suggested in the next part of this chapter.

The limitation of the MUSIG activation function as regards to its reliance on fixed and heuristic class responses has been mentioned in Chap. 3. It has been observed that a set of class boundaries may generate good segmentation in respect of certain criteria though it may or may not be guaranteed that the same set of class boundaries will give a good segmentation based on another criteria. A multi-objective genetic algorithm-based OptiMUSIG activation function for multilevel greyscale image segmentation is presented. This function is characterised by optimised class boundaries by incorporating the intensity gamut immanent in the input images. Different image segmentation quality measures are applied to derive the optimised class boundaries in this procedure. The selection of a particular set of class boundaries is performed based on another quantitative measure applied in the network. The performance of the multi-objective genetic algorithm-based OptiMUSIG activation function for the segmentation of real-life greyscale images shows superior performance in most of the cases as compared to the conventional MUSIG activation function with heuristic class boundaries.

One of the most efficient and popular elitist multi-objective genetic algorithm-based algorithm, NSGA II, is applied to generate class boundaries. The NSGA-II-based OptiMUSIG activation function for multilevel greyscale image segmentation is purported in the second part of this chapter. The optimised class boundaries those are evolved by NSGA-II are applied to generate the NSGA-II-based OptiMUSIG activation function. Different image segmentation quality measures are also used to derive the optimised class boundaries in this procedure. Another quantitative quality measure is applied to select a particular set of class boundaries that will be applied in the MLSONN network. The performance of the NSGA-II-based OptiMUSIG activation function for the segmentation of real-life greyscale images indicates superior performance in most of the cases as compared to the conventional MUSIG activation function with heuristic class boundaries.

The next chapter intends to overcome the drawback of single objective-based ParaOptiMUSIG activation function for colour image segmentation. The optimised class levels employed to generate the ParaOptiMUSIG activation function may be evaluated by a multi-objective optimisation procedure.

Chapter 6
Self-supervised Colour Image Segmentation Using Multiobjective Based Parallel Optimized MUSIG (ParaOptiMUSIG) Activation Function

6.1 Introduction

In Chap. 4, it has been illustrated and proved that the colour images are efficiently segmented by the ParaOptiMUSIG [193, 258, 273] activation function in connection with the parallel self-organizing neural network (PSONN) [195, 196] architecture. Individual objective functions are employed to generate the optimized class levels by the proposed approach in that chapter. Like the single objective based OptiMUSIG activation function in the previous chapter, it has been quite clear that the single objective based ParaOptiMUSIG activation function may not generate good colour segmented output image in respect of all other fitness criteria. The optimized class levels are generated on the basis of more than one criterion to overcome the problem. The derived class levels are efficient for all functional objectives. The problem can be treated as a multiobjective optimization problem and the solution can be found in that way. The researchers in various fields are attracted significantly by the evolutionary multiobjective optimization techniques due to their effectiveness and robustness in searching a set of trade-off solutions. The definition and working principle of the multiobjective optimization are already discussed in Chap. 5. Different types of segmentation/clustering using multiobjective criteria are also discussed in the previous chapter and the interested readers can also go through these articles [288–290]. An improved version of the NSGA-II has been applied [291] to solve the segmentation problem by changing the stopping criterion and automating the choice of the size of the initial population and the number of generations. Bandyopadhyay et al. [292] proposed a multiobjective genetic algorithm based classifier, named CEMOGA-classifier, for segmenting the remote sensing images. This classifier is be able to recognize the pixels belonging to a class given their intensity values in multiple bands.

This chapter is divided into two parts. In the first part, the true colour images are segmented into different number of classes with the optimized class levels for distinct colour components which are generated in parallel by means of multiobjective

S. De et al., *Hybrid Soft Computing for Multilevel Image and Data Segmentation*, Computational Intelligence Methods and Applications, DOI 10.1007/978-3-319-47524-0_6

genetic algorithm based optimization techniques. The distinct colour components are transferred in individual SONNs in a PSONN architecture after segregating the input true colour image into different colour components. Different activation functions are applied for different SONNs. The multiobjective genetic algorithm based optimized class boundaries are applied to design the parallel version of the optimized MUSIG (ParaOptiMUSIG) activation function for the PSONN [99]. The application of the multiobjective genetic algorithm based ParaOptiMUSIG activation function approach [99] is presented using true colour version of the Lena and Baboon images and a colour image of a human brain using a magnetic resonance imaging (MRI) machine [36]. The multiobjective genetic algorithm is employed to optimize three measures, viz., the standard measure of correlation coefficient (ρ) [72], F due to Liu and Yang [255] and F' due to Borsotti et al. [256]. Results of segmentation using the multiobjective genetic algorithm based ParaOptiMUSIG activation functions show better performance over the conventional MUSIG activation function employing heuristic class responses.

In next part of the chapter, the optimized class levels for distinct colour components which are generated in parallel by means of NSGA II based optimization techniques are applied to generate the ParaOptiMUSIG activation function [287, 293]. The individual SONNs in a PSONN architecture are fed with the distinct colour components after segregating the input true colour images into different colour components. The NSGA II based ParaOptiMUSIG activation function [287, 293] is employed in PSONN architecture to segment the colour images. The application of the NSGA II based ParaOptiMUSIG activation function [287, 293] approach is presented using two real life colour images, viz., the Baboon and Peppers images and the colour Brain MRI image [36]. The same fitness functions of the previous method are used as the fitness functions to be optimized in this NSGA II based method. Ultimately, it has been proved on the basis of the results that the segmentation using the NSGA II based ParaOptiMUSIG activation functions performs better than the conventional MUSIG activation function employing heuristic class responses.

The outline of this chapter is as follows. The next section is followed by the detailed representation of the multiobjective genetic algorithm based ParaOptiMUSIG activation function generation methodology to segment the colour images. The image segmentation results using this presented method, both in qualitative and quantitative nature, are detailed in the next section. The NSGA II based ParaOptiMUSIG activation function is described in the next section and this section is followed by the colour image segmentation procedure by the NSGA II based ParaOptiMUSIG activation function. After that, both the qualitative and quantitative comparisons between the image segmentation by the narrated method and the MUSIG activation function are demonstrated. The chapter ends with a brief concluding section.

6.2 Colour Image Segmentation by a Multiobjective Genetic Algorithm Based ParaOptiMUSIG Activation Function

The colour image segmentation by a multiobjective genetic algorithm based ParaOptiMUSIG activation function with a PSONN architecture has been demonstrated in this section and Fig. 6.1 depicts the flow diagram of this approach. This section is divided into two subsections. The methodology is described in the first part of this section and the second part of this section shows the experimental result of the methodology [99].

6.2.1 Methodology

This part of the chapter concentrated to segment the colour image by the multiobjective based ParaOptiMUSIG activation function with a PSONN architecture. This

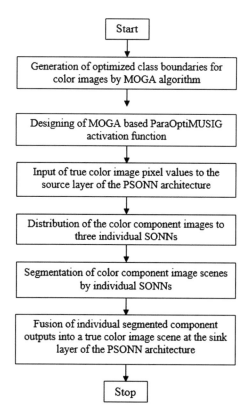

Fig. 6.1 Flow diagram of colour image segmentation using multiobjective genetic algorithm based ParaOptiMUSIG activation function

approach has been described into six phases. The following subsections elaborate the different phases of this approach [99].

6.2.1.1   Generation of Optimized Class Boundaries for True Colour Images

Like the generation of optimized class boundaries for grayscale images in the previous chapter, this is also an important part of the true colour image segmentation. The optimized class boundaries ($c_{\gamma_{opt}}$) are also generated by the multiobjective GA-based optimization procedure. These optimized class boundaries are applied to generate the ParaOptiMUSIG activation function. The procedure applied in this phase is detailed as follows.

- *Initialization phase*: The number of classes (K) and the true colour image pixel intensity levels are provided as inputs to this optimization procedure in this phase.
- *Chromosome representation and population generation*: The optimized class boundaries from the input true colour image information content are generated by the binary encoding technique for the chromosomes. Each pixel intensity of the true colour image information is differentiated into three colour components, viz., red, green and blue colour components. Three different chromosome pools are produced for the three individual colour components. Each chromosome pool is used to generate the optimized class levels for the individual colour component. A population size of 200 has been applied for this treatment.
- *Fitness computation*: Like the grayscale image segmentation, three segmentation efficiency measures (ρ, F, F') given in Eqs. 3.25, 3.28 and 3.29, respectively, are employed as the fitness function for this phase. These functions are applied as the evaluation functions in the multiobjective genetic algorithm. These fitness functions are applied on three chromosome pools in cumulative fashion.
- *Selection*: The non-dominated solutions are selected from the chromosome pool in this phase. These selected chromosomes are applied to generate the mating pool for the crossover and mutation.
- *Crossover*: A single point crossover operation is applied to generate the new pool of chromosomes. The crossover operation is utilized in red, green and blue chromosome pools separately. In this approach, the crossover probability is equal to 0.8.
- *Mutation*: The mutation operation is employed in red, green and blue chromosome pools separately. The mutation probability is taken as 0.01 in this approach.
- *New generation*: After mutation, the child chromosomes are combined with previously selected non-dominated chromosomes to make the new generation. This new population is again ready for new generation.

After a certain amount of iterations, the Pareto-optimal set of chromosomes is acquired. In this approach, this process continued for 100 iterations. This Pareto-optimal set is applied to generate the ParaOptiMUSIG activation function.

- *Selecting a solution from the non-dominated set*: A particular solution is selected from the non-dominated set of solutions after the final generation by the minimum valued Q index [256]. It is same as the grayscale image segmentation by multi-objective genetic algorithm based OptiMUSIG activation function in the previous chapter.

6.2.1.2 Designing Of Multiobjective Genetic Algorithm Based ParaOptiMUSIG Activation Function

The multiobjective genetic algorithm based ParaOptiMUSIG activation function is designed by the optimized class boundaries $(c_{\gamma_{opt}})$ those are selected in the Pareto-optimal set. The optimized class boundaries for individual colour component in the selected chromosomes are applied to generate the individual OptiMUSIG activation function for that colour component, viz., the class boundaries for the red component is employed to generate OptiMUSIG activation function for red and so on. After that the ParaOptiMUSIG activation function is generated.

6.2.1.3 Input of True Colour Image Pixel Values to the Source Layer of the PSONN Architecture

The pixel intensity levels of the true colour image, in this phase, are provided as inputs to the source layer of the PSONN architecture. The input image pixel true colour intensities are assigned to each of the neurons of the source layer for this purpose.

6.2.1.4 Distribution of the Colour Component Images to Three Individual SONNs

The individual primary colour components are separated from the pixel intensity levels of the input true colour image and the three individual three-layer component SONNs are inputted by these individual primary colour components, viz. the red component is applied to one SONN, the green component to another SONN and the remaining SONN accepts the blue component information at their respective input layers. The fixed interconnections of the respective SONNs with the source layer are responsible for this scenario.

6.2.1.5 Segmentation of Colour Component Images by Individual SONNs

The individual colour component of the true colour images is segmented by the corresponding SONN architecture guided by the designed multiobjective genetic algorithm based ParaOptiMUSIG activation function at the constituent primitives/ neurons. The neurons of different layers of individual three-layer SONN architecture generate different input colour component level responses, depending on the number of transition lobes of the ParaOptiMUSIG activation function. Since the network has no a priori knowledge about the outputs, the subnormal linear index of fuzziness [72] is applied to determine the system errors at the corresponding output layers. Like the grayscale image segmentation in the previous chapter, these errors are employed to adjust the interconnection weights between the different layers using the standard backpropagation algorithm [37, 47]. The respective output layers of the independent SONNs generate the final colour component images when the self-supervision of the corresponding networks attain stabilization.

6.2.1.6   Fusion of Individual Segmented Component Outputs into a True Colour Image at the Sink Layer of the PSONN Architecture

The finally segmented true colour image is derived at the sink layer of the PSONN architecture by combining the segmented outputs derived at the three output layers of the three independent three-layer SONN architectures. The number of segments is a combination of the number of transition lobes of the designed multiobjective genetic algorithm based ParaOptiMUSIG activation functions used during component level segmentation.

6.2.2   Experimental Results

The colour version of the two real-life images, viz., Lena and Baboon each of dimension 128×128 and a colour image of a human brain using a magnetic resonance imaging (MRI) machine [36] of dimension 170×170 are applied to describe the functionality of the multilevel true colour image segmentation approach using the PSONN architecture guided by the multiobjective genetic algorithm based ParaOptiMUSIG activation function. Experiments have been conducted with $K = \{6, 8\}$ classes. The ParaOptiMUSIG activation function has been designed with a fixed slope, $\lambda = \{2, 4\}$. Results are reported for 8 classes and a fixed slope of $\lambda = 4$. The segmentation effectivity of the present approach in accordance with the ParaOptiMUSIG activation function is shown in the following subsection. The present approach has been compared with the segmentation derived by means of the conventional MUSIG activation function with same number of class responses and heuristic class levels. The quantitative comparison of the above-mentioned methods is demonstrated in this subsection. In the next subsection, the comparative study as regards to its efficacy in the segmentation of test images of the segmented outputs by the present methods and the conventional MUSIG activation function based methods is illustrated.

6.2.2.1 Quantitative Performance Analysis of Segmentation

In this section, the quantitative measures of the efficiency of the multiobjective genetic algorithm based ParaOptiMUSIG activation function and the conventional MUSIG activation functions for $K = 8$ has been elaborated. Like the grayscale image segmentation by MOGA-based OptiMUSIG activation function in the previous chapter, same three segmentation efficiency functions, viz., correlation coefficient (ρ), and two empirical evaluation functions (F and F') have been employed in this multiobjective genetic algorithm based approaches to generate the activation functions and another empirical evaluation function (Q) has been used for the selection of the chromosome that is applied in the PSONN architecture network in this method. The results obtained with the ParaOptiMUSIG activation function have been illustrated in the following subsection and the corresponding results derived with the conventional fixed class response based MUSIG activation function are rendered in the next subsection.

1. *Segmentation evaluation using the multiobjective genetic algorithm based ParaOptiMUSIG activation function for colour images*: The optimized set of class boundaries for the colour images obtained using multiobjective genetic algorithm based algorithm with the three evaluation functions (ρ, F and F') is depicted in Tables 6.1, 6.2 and 6.3. These evaluation functions are applied as the fitness functions in the multiobjective genetic algorithm based procedure. The optimized set of class boundaries, derived as the Pareto-optimal set by the present approach and the corresponding three evaluation function (ρ, F and F') values of individual chromosome set are also reported in these tables. Four Pareto-optimal sets are reported for Lena and Baboon images and two Pareto-optimal sets are tabulated for the colour Brain MRI image. The Q values of individual set of class boundaries for colour images are presented in Tables 6.4, 6.5 and 6.6. The quality measures κ [graded on a scale of 1 (best) onwards] obtained by the segmentation of the test images based on the corresponding set of optimized class boundaries are shown in the last columns of the Tables 6.1, 6.2 and 6.3. The gradation has been performed on the basis of the Q values of the particular set of class boundaries. The first four good Q values obtained are indicated in those tables for easy reckoning. All evaluation function values are tabulated in normalized form in this chapter.
2. *MUSIG guided segmentation evaluation*: In Table 6.7, the heuristically selected class boundaries with the conventional MUSIG activation function for colour images are shown. The same evaluation functions are applied and the results are reported for the individual set of class boundaries. The corresponding Q values evaluated after the segmentation process are tabulated in Table 6.8. It is quite evident from Tables 6.4, 6.5, 6.6 and 6.8 that the fitness values derived by the ParaOptiMUSIG activation function for colour images are better than those obtained by the conventional MUSIG activation function.

The pareto-optimal set of optimized class boundaries and corresponding evaluated segmentation quality measures (ρ, F, F') for 8 classes of colour Lena image are

Table 6.1 Pareto-optimal set of optimized class boundaries and corresponding evaluated segmentation quality measures for 8 classes of colour Lena image

Set no		Class level	ρ	F	F'	κ
1	(i)	R = {26, 85, 102, 162, 176, 191, 211, 255}; G = {0, 17, 40, 55, 164, 195, 211, 255}; B = {21, 63, 108, 128, 146, 170, 185, 234}	0.95072	0.8096	0.6243	3
	(ii)	R = {26, 98, 112, 128, 152, 173, 194, 255}; G = {0, 19, 90, 133, 171, 195, 211, 255}; B = {21, 63, 82, 98, 123, 155, 189, 234}	0.9754	0.9636	0.8093	
	(iii)	R = {26, 88, 108, 141, 176, 209, 226, 255}; G = {0, 17, 70, 132, 167, 195, 211, 255}; B = {21, 63, 84, 104, 122, 146, 176, 234}	0.9754	0.9148	0.7249	
	(iv)	R = {26, 85, 102, 162, 177, 191, 211, 255}; G = {0, 21, 55, 69, 164, 195, 211, 255}; B = {21, 63, 82, 98, 128, 153, 191, 234}	0.9720	0.9122	0.7031	2
	(v)	R = {26, 87, 103, 127, 152, 189, 216, 255}; G = {0, 19, 88, 133, 151, 195, 211, 255}; B = {21, 63, 81, 98, 123, 146, 183, 234}	0.9753	0.9154	0.7049	4
	(vi)	R = {26, 85, 115, 146, 177, 201, 224, 255}; G = {0, 25, 114, 145, 164, 181, 195, 255}; B = {21, 63, 86, 110, 136, 167, 191, 234}	0.9698	0.8461	0.9088	1
	(vii)	R = {26, 85, 115, 147, 177, 200, 226, 255}; G = {0, 17, 50, 74, 132, 167, 195, 255}; B = {21, 66, 84, 108, 122, 146, 176, 234}	0.9688	1.0000	0.6750	
	(viii)	R = {26, 88, 108, 141, 178, 209, 226, 255}; G = {0, 41, 70, 145, 164, 195, 211, 255}; B = {21, 63, 79, 95, 136, 167, 191, 234}	0.9552	0.8377	0.6627	
	(ix)	R = {26, 88, 162, 177, 191, 216, 237, 255}; G = {0, 19, 88, 149, 170, 195, 211, 255}; B = {21, 63, 82, 109, 136, 173, 191, 234}	0.9643	0.8500	0.8842	
	(x)	R = {26, 85, 99, 115, 137, 177, 224, 255}; G = {0, 17, 90, 114, 164, 181, 195, 255}; B = {21, 63, 86, 102, 136, 167, 191, 234}	0.9622	0.8837	0.6505	

(continued)

Table 6.1 (continued)

Set no		Class level	ρ	F	F'	κ
2	(i)	R = {26, 85, 115, 151, 175, 221, 240, 255}; G = {0, 18, 32, 106, 164, 183, 211, 255}; B = {21, 64, 84, 111, 153, 167, 191, 234}	0.9621	0.7136	0.5874	
	(ii)	R = {26, 85, 115, 175, 189, 224, 239, 255}; G = {0, 18, 42, 133, 155, 195, 211, 255}; B = {21, 64, 85, 105, 127, 158, 191, 234}	0.9762	0.8116	0.5931	4
	(iii)	R = {26, 85, 113, 151, 166, 221, 240, 255}; G = {0, 18, 32, 54, 106, 164, 179, 255}; B = {21, 64, 79, 111, 133, 153, 171, 234}	0.9654	0.8027	0.6158	
	(iv)	R = {26, 85, 115, 151, 175, 213, 240, 255}; G = {0, 18, 90, 110, 164, 183, 211, 255}; B = {21, 64, 81, 105, 127, 158, 191, 234}	0.9757	0.7782	0.6386	1
	(v)	R = {26, 85, 115, 138, 189, 221, 240, 255}; G = {0, 18, 32, 106, 155, 197, 211, 255}; B = {21, 64, 93, 111, 153, 167, 189, 234}	0.9610	0.7202	0.5051	3
	(vi)	R = {26, 85, 115, 132, 151, 224, 238, 255}; G = {0, 18, 70, 106, 165, 183, 211, 255}; B = {21, 64, 79, 111, 137, 171, 190, 234}	0.9653	0.7618	0.5097	
	(vii)	R = {26, 85, 115, 143, 175, 215, 240, 255}; G = {0, 18, 72, 106, 120, 164, 183, 255}; B = {21, 64, 111, 140, 158, 174, 191, 234}	0.9399	0.7149	0.5501	
	(viii)	R = {26, 85, 115, 143, 175, 213, 240, 255}; G = {0, 18, 72, 107, 164, 183, 211, 255}; B = {21, 64, 111, 140, 158, 172, 191, 234}	0.9390	0.6704	0.4918	
	(ix)	R = {26, 85, 113, 159, 197, 221, 240, 255}; G = {0, 18, 32, 54, 106, 166, 183, 255}; B = {21, 64, 79, 105, 127, 155, 171, 234}	0.9691	0.8247	0.5602	2
	(x)	R = {26, 85, 115, 151, 175, 209, 240, 255}; G = {0, 18, 90, 107, 164, 179, 211, 255}; B = {21, 64, 79, 111, 144, 159, 182, 234}	0.9593	0.7159	0.5556	
	(xi)	R = {26, 85, 115, 142, 175, 215, 240, 255}; G = {0, 18, 72, 107, 160, 183, 208, 255}; B = {21, 64, 111, 141, 158, 174, 191, 234}	0.9398	0.6900	0.5349	
	(xii)	R = {26, 85, 113, 159, 175, 205, 240, 255}; G = {0, 18, 72, 106, 136, 164, 183, 255}; B = {21, 64, 100, 127, 158, 174, 191, 234}	0.9618	0.7312	0.5547	

(continued)

Table 6.1 (continued)

Set no		Class level	ρ	F	F'	κ
3	(i)	R = {26, 90, 106, 151, 175, 209, 233, 255}; G = {0, 18, 38, 137, 165, 190, 206, 255}; B = {21, 64, 80, 103, 126, 152, 189, 234}	0.9744	0.8906	0.8405	4
	(ii)	R = {26, 92, 109, 128, 151, 175, 209, 255}; G = {0, 18, 38, 137, 158, 181, 204, 255}; B = {21, 64, 81, 103, 126, 166, 189, 234}	0.9728	0.9195	0.6521	
	(iii)	R = {26, 92, 109, 145, 175, 204, 225, 255}; G = {0, 18, 64, 136, 166, 181, 204, 255}; B = {21, 64, 82, 102, 127, 153, 185, 234}	0.9751	0.9081	0.7263	1
	(iv)	R = {26, 92, 109, 138, 175, 204, 225, 255}; G = {0, 18, 64, 140, 166, 181, 207, 255}; B = {21, 64, 82, 102, 125, 153, 185, 234}	0.9758	0.9213	0.6876	2
	(v)	R = {26, 92, 109, 138, 175, 199, 227, 255}; G = {0, 18, 68, 140, 166, 181, 207, 255}; B = {21, 64, 82, 102, 125, 153, 180, 234}	0.9756	0.9114	0.8041	3
	(vi)	R = {26, 92, 107, 142, 175, 204, 237, 255}; G = {0, 22, 39, 78, 136, 178, 202, 255}; B = {21, 64, 102, 127, 153, 169, 184, 234}	0.9569	0.8546	0.6675	
4	(i)	R = {26, 84, 127, 146, 172, 197, 240, 255}; G = {0, 17, 44, 71, 122, 167, 210, 255}; B = {21, 63, 76, 94, 118, 151, 189, 234}	0.9738	0.8678	1.0000	2
	(ii)	R = {26, 84, 127, 144, 196, 227, 240, 255}; G = {0, 17, 37, 67, 122, 191, 210, 255}; B = {21, 64, 94, 118, 143, 162, 183, 234}	0.9702	0.7561	0.6840	4
	(iii)	R = {26, 94, 112, 150, 180, 213, 237, 255}; G = {0, 19, 55, 147, 162, 182, 208, 255}; B = {21, 63, 81, 102, 126, 151, 182, 234}	0.9766	0.8988	0.8045	3
	(iv)	R = {26, 84, 125, 147, 197, 227, 240, 255}; G = {0, 17, 39, 99, 122, 191, 210, 255}; B = {21, 63, 78, 94, 118, 151, 175, 234}	0.9737	0.8493	0.7783	1

Table 6.2 Pareto-optimal set of optimized class boundaries and corresponding evaluated segmentation quality measures for 8 classes of colour Baboon image

Set no	Class level	ρ	F	F'	κ
1	(i) R = {0, 41, 58, 84, 127, 157, 234, 255}; G = {0, 38, 63, 109, 134, 172, 190, 246}; B = {0, 40, 70, 101, 133, 172, 220, 255}	0.9832	0.7258	0.5043	1
	(ii) R = {0, 41, 58, 127, 157, 212, 234, 255}; G = {0, 38, 63, 108, 134, 174, 190, 246}; B = {0, 20, 70, 84, 133, 172, 228, 255}	0.9749	0.7042	0.3912	
	(iii) R = {0, 41, 58, 85, 127, 155, 230, 255}; G = {0, 38, 62, 112, 161, 176, 190, 246}; B = {0, 44, 89, 140, 169, 208, 229, 255}	0.9714	0.6589	0.4772	
	(iv) R = {0, 41, 58, 76, 127, 159, 238, 255}; G = {0, 38, 62, 131, 155, 174, 190, 246}; B = {0, 24, 84, 132, 153, 168, 208, 255}	0.9644	0.6314	0.4639	
	(v) R = {0, 42, 58, 85, 127, 158, 230, 255}; G = {0, 38, 62, 104, 126, 175, 193, 246}; B = {0, 20, 88, 132, 168, 208, 228, 255}	0.9622	0.5348	0.3809	
	(vi) R = {0, 41, 58, 84, 127, 156, 234, 255}; G = {0, 38, 63, 105, 135, 166, 190, 246}; B = {0, 20, 70, 84, 133, 188, 228, 255}	0.9722	0.6945	0.4899	4
	(vii) R = {0, 41, 58, 85, 127, 155, 202, 255}; G = {0, 38, 62, 107, 129, 176, 191, 246}; B = {0, 44, 89, 135, 169, 208, 229, 255}	0.9737	0.6777	0.5666	
	(viii) R = {0, 41, 58, 84, 119, 157, 234, 255}; G = {0, 38, 63, 99, 124, 166, 190, 246}; B = {0, 40, 70, 100, 133, 172, 220, 255}	0.9832	0.7369	0.6335	2
	(ix) R = {0, 58, 85, 127, 159, 173, 238, 255}; G = {0, 38, 62, 112, 141, 161, 190, 246}; B = {0, 24, 84, 132, 153, 208, 232, 255}	0.9661	0.6295	0.5042	
	(x) R = {0, 41, 58, 84, 127, 158, 231, 255}; G = {0, 38, 69, 91, 126, 169, 193, 246}; B = {0, 20, 70, 84, 132, 172, 228, 255}	0.9751	0.7272	0.4475	3
2	(i) R = {0, 44, 65, 93, 167, 182, 198, 255}; G = {0, 40, 55, 72, 146, 174, 190, 246}; B = {0, 28, 71, 104, 123, 152, 219, 255}	0.9801	0.7426	0.4776	
	(ii) R = {0, 44, 64, 93, 125, 166, 206, 255}; G = {0, 41, 55, 72, 125, 174, 190, 246}; B = {0, 20, 77, 129, 152, 179, 232, 255}	0.9695	0.5879	0.3222	

(continued)

Table 6.2 (continued)

Set no	Class level	ρ	F	F'	κ
(iii)	R = {0, 44, 65, 125, 182, 198, 238, 255}; G = {0, 40, 55, 74, 146, 174, 190, 246}; B = {0, 28, 81, 104, 127, 166, 219, 255}	0.9772	0.6853	0.4685	
(iv)	R = {0, 44, 77, 99, 192, 214, 230, 255}; G = {0, 41, 55, 72, 157, 172, 188, 246}; B = {0, 20, 72, 119, 144, 181, 232, 255}	0.9745	0.6143	0.4069	
(v)	R = {0, 44, 79, 99, 192, 214, 230, 255}; G = {0, 41, 55, 72, 140, 159, 188, 246}; B = {0, 20, 72, 119, 144, 183, 232, 255}	0.9742	0.5963	0.3633	
(vi)	R = {0, 44, 64, 125, 182, 198, 221, 255}; G = {0, 41, 55, 72, 128, 157, 174, 246}; B = {0, 28, 66, 104, 127, 160, 219, 255}	0.9806	0.7463	0.6875	
(vii)	R = {0, 44, 64, 93, 118, 166, 230, 255}; G = {0, 40, 55, 72, 141, 170, 190, 246}; B = {0, 28, 66, 105, 127, 169, 219, 255}	0.9799	0.7054	0.4932	3
(viii)	R = {0, 45, 67, 166, 206, 221, 235, 255}; G = {0, 41, 55, 72, 125, 172, 190, 246}; B = {0, 20, 77, 128, 179, 216, 232, 255}	0.9653	0.5126	0.2916	
(ix)	R = {0, 44, 67, 102, 125, 192, 230, 255}; G = {0, 41, 55, 72, 132, 159, 188, 246}; B = {0, 20, 72, 115, 144, 183, 232, 255}	0.9757	0.6504	0.4399	4
(x)	R = {0, 44, 77, 99, 135, 149, 206, 255}; G = {0, 41, 72, 119, 157, 172, 188, 246}; B = {0, 45, 72, 97, 127, 155, 225, 255}	0.9836	0.8994	0.6527	1
(xi)	R = {0, 44, 66, 102, 125, 192, 230, 255}; G = {0, 41, 55, 72, 132, 172, 188, 246}; B = {0, 20, 72, 115, 144, 183, 232, 255}	0.9758	0.6294	0.4699	
(xii)	R = {0, 44, 65, 93, 151, 183, 206, 255}; G = {0, 40, 55, 72, 138, 158, 174, 246}; B = {0, 28, 65, 97, 127, 157, 217, 255}	0.9816	0.7926	0.7055	
(xiii)	R = {0, 46, 74, 106, 130, 182, 238, 255}; G = {0, 41, 55, 74, 124, 146, 190, 246}; B = {0, 60, 81, 104, 127, 166, 219, 255}	0.9824	0.8515	0.6715	2
(xiv)	R = {0, 44, 61, 97, 166, 187, 230, 255}; G = {0, 40, 55, 88, 141, 175, 190, 246}; B = {0, 28, 71, 92, 120, 166, 223, 255}	0.9808	0.7962	0.5465	

(continued)

Table 6.2 (continued)

Set no		Class level	ρ	F	F'	κ
	(xv)	R = {0, 45, 65, 125, 166, 207, 238, 255}; G = {0, 40, 55, 74, 125, 175, 190, 246}; B = {0, 20, 75, 92, 134, 179, 223, 255}	0.9753	0.6506	0.4279	
	(xvi)	R = {0, 45, 67, 172, 206, 221, 235, 255}; G = {0, 41, 55, 72, 127, 142, 190, 255}; B = {0, 20, 111, 152, 179, 200, 216, 255}	0.9200	0.4865	0.4264	
3	(i)	R = {0, 45, 71, 123, 175, 191, 218, 255}; G = {0, 40, 59, 77, 137, 164, 189, 246}; B = {0, 41, 69, 96, 129, 170, 219, 255}	0.9830	0.8439	0.5481	4
	(ii)	R = {0, 49, 72, 161, 181, 206, 224, 255}; G = {0, 52, 69, 107, 125, 168, 188, 246}; B = {0, 21, 91,107, 150, 197, 232, 255}	0.9609	0.6551	0.4388	
	(iii)	R = {0, 45, 71, 117, 154, 191, 225, 255}; G = {0, 40, 59, 77, 137, 173, 191, 246}; B = {0, 21, 107, 150, 171, 197, 223, 255}	0.9321	0.5928	0.4395	
	(iv)	R = {0, 49, 101, 137, 165, 200, 227, 255}; G = {0, 40, 68, 85, 105, 171, 188, 246}; B = {0, 45, 69, 97, 129, 169, 228, 255}	0.9833	0.9533	0.7805	2
	(v)	R = {0, 45, 71, 88, 104, 191, 219, 255}; G = {0, 59, 77, 137, 155, 172, 190, 246}; B = {0, 21, 107, 147, 170, 197, 223, 255}	0.9327	0.6132	0.3464	
	(vi)	R = {0, 45, 71, 117, 148, 191, 225, 255}; G = {0, 43, 65, 117, 135, 159, 181, 246}; B = {0, 41, 85, 114, 150, 181, 219, 255}	0.9794	0.7678	0.5974	1
	(vii)	R = {0, 49, 73, 137, 161, 184, 224, 255}; G = {0, 40, 59, 88, 131, 166, 188, 246}; B = {0, 33, 69, 115, 155, 177, 223, 255}	0.9768	0.7294	0.5407	
	(viii)	R = {0, 49, 73, 89, 149, 165, 226, 255}; G = {0, 40, 57, 73, 137, 164, 188, 246}; B = {0, 36, 71, 115, 155, 204, 223, 255}	0.9768	0.7101	0.5329	3
	(ix)	R = {0, 49, 91, 161, 181, 200, 224, 255}; G = {0, 40, 59, 77, 105, 173, 191, 246}; B = {0, 21, 91, 107, 150, 171, 213, 255}	0.9597	0.6617	0.3947	
4	(i)	R = {0, 44, 108, 128, 170, 187, 206, 255}; G = {0, 44, 77, 97, 113, 159, 179, 246}; B = {0, 49,79, 110, 131, 176, 223, 255}	0.9825	0.8712	0.9039	2
	(ii)	R = {0, 45, 74, 116, 142, 211, 236, 255}; G = {0, 41, 70, 86, 120, 151, 187, 246}; B = {0, 49, 88, 117, 154, 177, 219, 255}	0.9801	0.8014	0.8528	

(continued)

Table 6.2 (continued)

Set no	Class level	ρ	F	F'	κ
(iii)	R = {0, 43, 77, 118, 142, 210, 236, 255}; G = {0, 45, 80, 97, 115, 149, 187, 246}; B = {0, 26, 79, 109, 149, 196, 223, 255}	0.9752	0.7175	0.5707	
(iv)	R = {0, 46, 74, 118, 173, 212, 236, 255}; G = {0, 44, 78, 99, 120, 157, 179, 246}; B = {0, 26, 76, 105, 126, 176, 223, 255}	0.9774	0.8120	0.6128	4
(v)	R = {0, 46, 70, 118, 147, 191, 236, 255}; G = {0, 40, 70, 98, 144, 164, 192, 246}; B = {0, 26, 76, 118, 176, 196, 223, 255}	0.9694	0.6816	0.5617	
(vi)	R = {0, 55, 99, 116, 140, 211, 236, 255}; G = {0, 70, 86, 120, 141, 159, 187, 246}; B = {0, 49, 79, 110, 131, 160, 223, 255}	0.9835	1.0000	1.0000	1
(vii)	R = {0, 45, 72, 140, 174, 204, 236, 255}; G = {0, 44, 70, 113, 138, 157, 179, 246}; B = {0, 26, 78, 116, 176, 192, 223, 255}	0.9686	0.6632	0.5470	
(viii)	R = {0, 45, 74, 128, 158, 187, 236, 255}; G = {0, 44, 70, 90, 113, 159, 179, 246}; B = {0, 49, 79, 106, 131, 176, 222, 255}	0.9831	0.9269	0.9910	3
(ix)	R = {0, 46, 62, 84, 118, 211, 231, 255}; G = {0, 44, 62, 83, 146, 162, 193, 246}; B = {0, 26, 76, 132, 176, 196, 223, 255}	0.9665	0.6411	0.5625	
(x)	R = {0, 44, 60, 85, 116, 142, 211, 255}; G = {0, 44, 61, 86, 109, 159, 187, 246}; B = {0, 49, 88, 110, 149, 177, 217, 255}	0.9799	0.8248	0.8313	
(xi)	R = {0, 46, 85, 102, 173, 211, 230, 255}; G = {0, 44, 61, 83, 109, 143, 193, 246}; B = {0, 49, 76, 110, 151, 177, 217, 255}	0.9811	0.8676	0.7537	
(xii)	R = {0, 43, 63, 118, 142, 210, 233, 255}; G = {0, 40, 64, 97, 115, 148, 179, 246}; B = {0, 45, 79, 117, 158, 197, 223, 255}	0.9796	0.8168	0.7799	
(xiii)	R = {0, 43, 77, 118, 142, 211, 236, 255}; G = {0, 44, 66, 97, 113, 135, 179, 246}; B = {0, 26, 88, 143, 179, 196, 219, 255}	0.9591	0.7103	0.4639	
(xiv)	R = {0, 46, 74, 92, 118, 174, 236, 255}; G = {0, 46, 70, 98, 138, 158, 179, 246}; B = {0, 26, 76, 114, 176, 196, 223, 255}	0.9682	0.7264	0.5457	
(xv)	R = {0, 46, 62, 84, 107, 132, 231, 255}; G = {0, 44, 67, 95, 113, 146, 179, 246}; B = {0, 21, 48, 79, 131, 147, 222, 255}	0.9718	0.7537	0.5004	

Table 6.3 Pareto-optimal set of optimized class boundaries and corresponding evaluated segmentation quality measures for 8 classes of colour Brain MRI image

Set no	Class level		ρ	F	F'	κ
1	(i)	R = {2, 17, 35, 94, 137, 166, 227, 254}; G = {3, 16, 32, 59, 88, 151, 183, 236}; B = {0, 18, 109, 123, 146, 199, 224, 254}	0.9862	0.6670	0.6439	3
	(ii)	R = {2, 17, 83, 111, 136, 168, 228, 254}; G = {3, 32, 59, 79, 113, 135, 214, 236}; B = {0, 66, 113, 146, 170, 206, 234, 254}	0.9912	0.7954	0.7853	
	(iii)	R = {2, 17, 51, 137, 167, 211, 231, 254}; G = {3, 16, 32, 58, 135, 183, 214, 236}; B = {0, 18, 66, 101, 146, 171, 234, 254}	0.9914	0.5966	0.6822	1
	(iv)	R = {2, 17, 35, 111, 174, 195, 227, 254}; G = {3, 16, 32, 59, 88, 150, 183, 236}; B = {0, 18, 66, 105, 146, 170, 234, 254}	0.9906	0.6186	0.5666	2
	(v)	R = {2, 17, 35, 79, 136, 172, 193, 254}; G = {3, 16, 32, 46, 88, 159, 183, 236}; B = {0, 18, 109, 146, 178, 193, 225, 254}	0.9850	0.6538	0.5194	
	(vi)	R = {2, 17, 51, 111, 135, 202, 231, 254}; G = {3, 17, 32, 59, 159, 183, 212, 236}; B = {0, 101, 146, 170, 186, 210, 225, 254}	0.9843	0.6586	0.6859	4
2	(i)	R = {2, 23, 40, 90, 148, 201, 235, 254}; G = {3, 17, 52, 113, 135, 158, 187, 236}; B = {0, 13, 72, 100, 138, 161, 175, 254}	0.9886	0.5907	0.5056	2
	(ii)	R = {2, 23, 90, 148, 168, 201, 235, 254}; G = {3, 17, 52, 113, 131, 158, 187, 236}; B = {0, 13, 33, 72, 100, 138, 175, 254}	0.9913	0.7207	0.6266	
	(iii)	R = {2, 23, 74, 148, 198, 216, 239, 254}; G = {3, 17, 52, 67, 122, 135, 190, 236}; B = {0, 48, 73, 100, 138, 175, 225, 254}	0.9907	0.7528	0.6059	
	(iv)	R = {2, 23, 72, 148, 201, 216, 239, 254}; G = {3, 17, 52, 67, 122, 152, 190, 236}; B = {0, 13, 73, 100, 138, 161, 239, 254}	0.9878	0.5224	0.4222	3
	(v)	R = {2, 22, 40, 90, 126, 212, 235, 254}; G = {3, 17, 52, 113, 135, 187, 220, 236}; B = {0, 13, 72, 104, 142, 173, 202, 254}	0.9889	0.5669	0.5105	1
	(vi)	R = { 2, 21, 72, 148, 202, 216, 239, 254}; G = {3, 17, 67, 104, 122, 152, 190, 254}; B = {0, 41, 73, 100, 138, 175, 225, 254}	0.9875	0.6886	0.5681	
	(vii)	R = {22, 90, 126, 168, 212, 235, 254}; G = {3, 25, 52, 113, 135, 187, 220, 236}; B = {0, 15, 72, 104, 142, 173, 202, 254}	0.9893	0.6634	0.7147	4

Table 6.4 The evaluated segmentation quality measure Q of the each optimized class boundaries of Table 6.1

Set	No	(i)	(ii)	(iii)	(iv)	(v)	(vi)	(vii)
1	Q	0.5624	0.7102	0.5877	0.5524	0.5701	0.5402	0.5889
	No	(viii)	(ix)	(x)				
	Q	0.6065	0.5887	0.6468				
2	No	(i)	(ii)	(iii)	(iv)	(v)	(vi)	(vii)
	Q	0.6984	0.6016	0.7643	0.5611	0.5970	0.9489	0.7182
	No	(viii)	(ix)	(x)	(xi)	(xii)		
	Q	0.9353	0.5928	0.6971	0.9404	0.6195		
3	No	(i)	(ii)	(iii)	(iv)	(v)	(vi)	
	Q	0.6400	0.8288	0.6182	0.6269	0.6377	0.6802	
4	No	(i)	(ii)	(iii)	(iv)			
	Q	0.5935	0.6359	0.6127	0.5935			

Table 6.5 The evaluated segmentation quality measure Q of the each optimized class boundaries of Table 6.2

Set	No	(i)	(ii)	(iii)	(iv)	(v)	(vi)	(vii)
1	Q	0.4723	0.5191	0.6752	0.6241	0.6225	0.5124	0.6096
	No	(viii)	(ix)	(x)				
	Q	0.5005	0.6028	0.5007				
2	No	(i)	(ii)	(iii)	(iv)	(v)	(vi)	(vii)
	Q	0.5826	0.5927	0.7038	0.7262	0.7097	0.6347	0.5078
	No	(viii)	(ix)	(x)	(xi)	(xii)	(xiii)	(xiv)
	Q	0.7929	0.5312	0.4207	0.5391	0.5303	0.4433	0.5591
	No	(xv)	(xvi)					
	Q	0.6279	0.9144					
3	No	(i)	(ii)	(iii)	(iv)	(v)	(vi)	(vii)
	Q	0.6141	0.6518	0.8596	0.5619	0.7925	0.4922	0.6523
	No	(viii)	(ix)					
	Q	0.5899	0.7733					
4	No	(i)	(ii)	(iii)	(iv)	(v)	(vi)	(vii)
	Q	0.4568	0.5325	0.4933	0.4660	0.5684	0.4311	0.5828
	No	(viii)	(ix)	(x)	(xi)	(xii)	(xiii)	(xiv)
	Q	0.4623	0.8016	0.5057	0.5645	0.5499	0.5774	0.5334
	No	(xv)						
	Q	0.5191						

Table 6.6 The evaluated segmentation quality measure Q of the each optimized class boundaries of Table 6.3

Set	No	(i)	(ii)	(iii)	(iv)	(v)	(vi)	
1	Q	0.3747	0.4673	0.3673	0.3685	0.3957	0.3878	
2	No	(i)	(ii)	(iii)	(iv)	(v)	(vi)	(vii)
	Q	0.3741	0.4140	0.4365	0.3853	0.3535	0.4091	0.3857

tabulated in Table 6.1 and the corresponding fixed class boundaries and the evaluated segmentation quality measures for 8 classes of the same image are shown in Table 6.7. Most of the cases, it is observed that the segmentation quality measures derived by the narrated method are better than the same deduced using the conventional fixed class responses. It is also detected that one of the segmentation quality measures obtained by the ParaOptiMUSIG activation function based method may not be better than the same found by the conventional MUSIG activation function. The same thing is also observed when we go though the Tables 6.2 and 6.3 for colour Baboon and colour Brain MRI images, respectively. In these tables, the pareto-optimal set of optimized class boundaries and corresponding evaluated segmentation quality measure (ρ, F, F') values are reported. The fixed class responses and the corresponding evaluated segmentation quality measure values are tabulated in Table 6.7 for colour baboon and colour Brain MRI image. The ParaOptiMUSIG activation function based method does better segmentation of the Baboon and Brain MRI images than the MUSIG activation function with respect to the three segmentation quality measures (ρ, F, F'). The evaluated segmentation quality measure Q of the each optimized class boundaries is tabulated in Tables 6.4, 6.5 and 6.6 for the test images and the same of the fixed class boundaries is reported in Table 6.8. A detailed scrutiny revealed that the Q values reported in Tables 6.4, 6.5 and 6.6 are much better that the same tabulated in Table 6.8. So, it can be concluded that the MOGA-based ParaOptiMUSIG activation function can do better colour image segmentation than the conventional MUSIC activation function.

6.2.3 Image Segmentation Outputs

The segmented true colour output images obtained for the $K = 8$ classes, with the narrated optimized approach vis-a-vis those obtained with the heuristically chosen class boundaries, are exhibited in this section.

Table 6.7 Fixed class boundaries and corresponding evaluated segmentation quality measures for 8 classes of colour images

Image	Set no	Class level	ρ	F	F'
Lena	1	R = {26, 80, 110, 130, 150, 220, 240, 255}; G = {0, 20, 70, 100, 160, 185, 210, 255}; B = {21, 60, 80, 110, 135, 170, 190, 234}	0.9673	0.8240	0.7438
	2	R = {26, 90, 100, 120, 165, 210, 235, 255}; G = {0, 30, 65, 90, 150, 180, 200, 255}; B = {21, 55, 75, 140, 155, 175, 190, 234}	0.9036	0.8006	0.6214
	3	R = {26, 80, 90, 115, 145, 210, 235, 255}; G = {0, 10, 55, 95, 150, 165, 190, 255}; B = {21, 50, 80, 140, 160, 170, 210, 234}	0.9045	0.6939	0.5798
	4	R = {26, 81, 130, 140, 170, 210, 240, 255}; G = {0, 30, 74, 120, 155, 200, 215, 255}; B = {21, 50, 60, 124, 160, 185, 200, 255}	0.9023	0.7012	0.6733
Baboon	1	R = {0, 45, 65, 170, 210, 220, 235, 255}; G = {0, 40, 55, 70, 125, 140, 190, 255}; B = {0, 20, 110, 150, 180, 200, 216, 255}	0.9226	0.5029	0.3438
	2	R = {0, 50, 70, 175, 205, 225, 230, 255}; G = {0, 35, 50, 65, 120, 135, 185, 255}; B = {0, 15, 105, 145, 175, 195, 210, 255}	0.9248	0.5064	0.3611
	3	R = {0, 40, 55, 105, 175, 220, 235, 255}; G = {0, 45, 65, 120, 150, 170, 195, 255}; B = {0, 25, 75, 165, 180, 215, 235, 255}	0.9358	0.5517	0.2788
	4	R = {0, 35, 60, 110, 180, 215, 225, 255}; G = {0, 40, 70, 100, 145, 175, 190, 255}; B = {0, 30, 95, 160, 182, 205, 230, 255}	0.9491	0.6637	0.3613
MRI	1	R = {2, 30, 60, 90, 120, 150, 180, 254}; G = {3, 10, 50, 90, 130, 170, 210, 236}; B = {0, 40, 60, 75, 140, 150, 200, 254}	0.9838	0.7692	0.6260
	2	R = {2, 25, 65, 105, 155, 205, 225, 254}; G = {3, 20, 50, 80, 135, 175, 215, 236}; B = {0, 35, 55, 75, 125, 155, 210, 254}	0.9872	0.7359	0.6891
	3	R = {2, 23, 70, 120, 160, 180, 205, 254}; G = {3, 22, 52, 75, 125, 165, 200, 236}; B = {0, 30, 65, 82, 135, 160, 195, 254}	0.9884	0.7342	0.6659
	4	R = {2, 28, 55, 125, 170, 190, 215, 254}; G = {3, 25, 60, 78, 122, 152, 205, 236}; B = {0, 45, 70, 100, 130, 170, 220, 254}	0.9848	1.0000	1.0000

Table 6.8 The evaluated segmentation quality measure Q of the each fixed class boundaries of Table 6.7

Image	Set no	Q	Image	Set no	Q	Image	Set no	Q
Lena	1	0.8675	Baboon	1	0.9173	MRI	1	1.0000
	2	0.8151		2	0.9340		2	0.5269
	3	0.8815		3	1.0000		3	0.5980
	4	1.0000		4	0.9009		4	0.5868

6.2.3.1   Multiobjective Genetic Algorithm Based ParaOptiMUSIG Guided Segmented Outputs

The first four better results of each Pareto-optimal set are applied in the networks. The colour test images derived with the PSONN architecture using the ParaOptiMUSIG activation function for $K = 8$ are shown in Figs. 6.2, 6.3 and 6.4.

6.2.3.2   MUSIG Guided Segmented Outputs

The segmented test colour images obtained with the PSONN architecture characterised by the conventional MUSIG activation employing fixed class responses for $K = 8$, are shown in Figs. 6.5 and 6.6.

From the results obtained, it is evident that the multiobjective genetic algorithm based OptiMUSIG activation function for colour images outperforms its conventional MUSIG counterpart as regards to the segmentation quality of the images for the different number of classes. Moreover, since the ParaOptiMUSIG activation function based approach incorporates the image heterogeneity, it can handle a wide variety of image intensity distribution prevalent in real life.

6.3   NSGA II Based Parallel Optimized Multilevel Sigmoidal (ParaOptiMUSIG) Activation Function

Like the NSGA II based OptiMUSIG activation function, the most popular multiobjective genetic algorithm, NSGA II, is also employed to generate the optimized class boundaries for the colour image segmentation. The NSGA-II based ParaOptiMUSIG activation function $(\overline{f}_{Para_{NSGA}}(\overline{x}))$ [99, 293] is presented in this article to segment the colour images on the basis of more than one criteria. This function $(\overline{f}_{Para_{NSGA}}(\overline{x}))$ is represented as [287, 293]

$$\overline{f}_{Para_{NSGA}}(\overline{x}) = [\overline{f}_{Para_{F_1}}(\overline{x}), \overline{f}_{Para_{F_2}}(\overline{x}), .., \overline{f}_{Para_{F_n}}(\overline{x})]^T \tag{6.1}$$

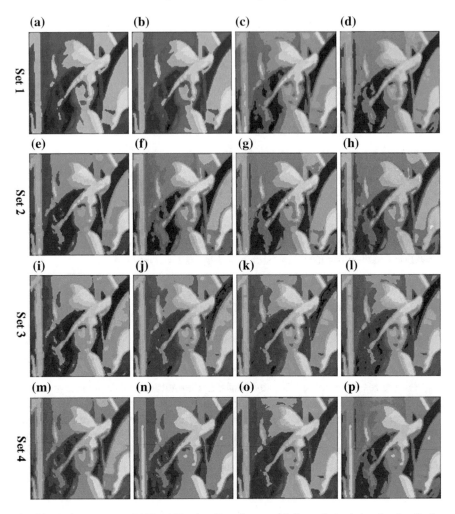

Fig. 6.2   8-class segmented 128×128 colour Lena image with the optimized class levels referring to **a–d** set 1 **e–h** set 2 **i–l** set 3 **m–p** set 4 of Table 6.1 for first four better quality measure Q with multiobjective genetic algorithm based ParaOptiMUSIG activation function

subject to $g_{Para_{F_i}}(\overline{x}) \geq 0, i = 1, 2, \ldots, m$ and $h_{Para_{F_i}}(\overline{x}) = 0, i = 1, 2, \ldots, p$ where, $f_{Para_{F_i}}(\overline{x}) \geq 0, i = 1, 2, \ldots, n$ are different objective functions in the NSGA-II based ParaOptiMUSIG activation function. $g_{Para_{F_i}}(\overline{x}) \geq 0$, $i = 1, 2, \ldots, m$ and $h_{Para_{F_i}}(\overline{x}) = 0, i = 1, 2, \ldots, p$ are the inequality and equality constraints of this function, respectively [99, 293].

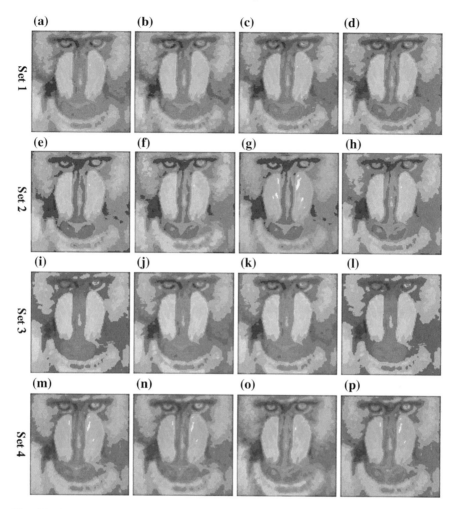

Fig. 6.3 8-class segmented 128×128 colour Baboon image with the optimized class levels referring to **a–d** set 1 **e–h** set 2 **i–l** set 3 **m–p** set 4 of Table 6.2 for first four better quality measure Q with multiobjective genetic algorithm based ParaOptiMUSIG activation function

6.3.1 Colour Image Segmentation By NSGA II Based ParaOptiMUSIG Activation Function

This methodology for colour image segmentation by the NSGA II based ParaOpti-MUSIG activation function with a PSONN architecture has been described in this section. The flow diagram of this method is shown in Fig. 6.7. The following sub-sections detailed the different phases of this approach [287].

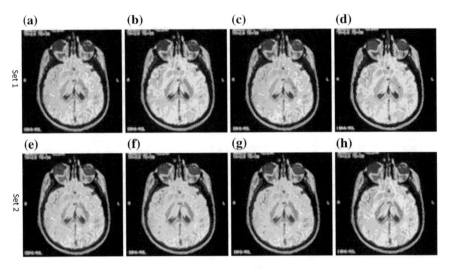

Fig. 6.4 8-class segmented 170×170 colour Brain MRI image with the optimized class levels referring to **a–d** set 1 **e–h** set 2 of Table 6.3 for first four better quality measure Q with multiobjective genetic algorithm based ParaOptiMUSIG activation function

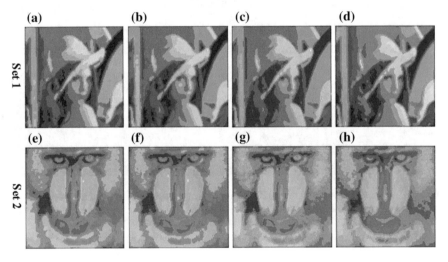

Fig. 6.5 8-class segmented 128×128 **a–d** colour Lena image and **e–h** colour Baboon image with the fixed class levels of Table 6.7 with the MUSIG activation function

6.3.1.1   Generation of Optimized Class Boundaries for Colour Images by NSGA II Algorithm

In this part of the chapter, the optimized class boundaries, like the grayscale images in Chap. 5, are generated for the segmentation of the colour images. The NSGA II based optimization procedure is employed to generate the optimized class boundaries

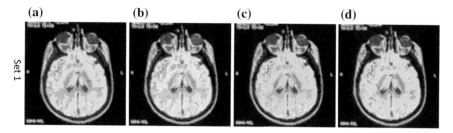

Set 1

(a) **(b)** **(c)** **(d)**

Fig. 6.6 8-class segmented 170×170 **a–d** colour Brain MRI image with the fixed class levels of Table 6.7 with the MUSIG activation function

Fig. 6.7 Flow diagram of colour image segmentation using NSGA II based ParaOptiMUSIG activation function

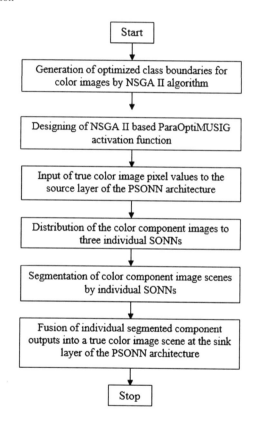

Start

↓

Generation of optimized class boundaries for color images by NSGA II algorithm

↓

Designing of NSGA II based ParaOptiMUSIG activation function

↓

Input of true color image pixel values to the source layer of the PSONN architecture

↓

Distribution of the color component images to three individual SONNs

↓

Segmentation of color component image scenes by individual SONNs

↓

Fusion of individual segmented component outputs into a true color image scene at the sink layer of the PSONN architecture

↓

Stop

$(c_{\gamma_{opt}})$ for the NSGA II based ParaOptiMUSIG activation function [287, 293]. The procedure applied in this phase is detailed as follows.

- *Input phase*: The pixel intensity levels of the colour image and the number of classes (K) to be segmented are furnished as inputs to this NSGA II based optimization procedure.
- *Chromosome representation and population generation*: The real coded chromosomes are applied in NSGA II based optimization procedure and the real numbers

are generated randomly for developing the optimized class boundaries from the input colour image information content. The three colour components, viz. red, green and blue colour components are differentiated from each pixel intensity of the true colour image information. Three different chromosome pools are developed for the three individual colour components and individual chromosome pool is employed to generate the optimized class levels for the individual colour component. In this treatment, a population size of 200 has been applied.

- *Fitness computation*: Three segmentation efficiency measures (ρ, F, F'), given in Eqs. 3.25, 3.28 and 3.29 respectively, are utilized as the evaluation functions in the NSGA II algorithm. These fitness functions are applied on three chromosome pools in cumulative fashion.
- *Genetic operators*: Same genetic operators, like grayscale image segmentation by NSGA II based OptiMUSIG activation function in the previous chapter, are also used is this phase. The crowded binary tournament selection operator is also employed in NSGA II algorithm. Like the grayscale segmentation, the crossover probability and the mutation probability are selected as 0.8 and 0.01, respectively. The non-dominated solutions among the parent and child populations are transmitted for the next generation of NSGA II. The near-pareto-optimal strings furnish the desired solutions in the last generation.
- *Selecting a solution from the non-dominated set*: Q index [256] is applied to select the particular solution from the Pareto-optimal non-dominated set of solutions after the final generation.

6.3.1.2 Designing Of NSGA II Based ParaOptiMUSIG Activation Function

The optimized class boundaries ($c_{\gamma_{opt}}$) selected in the Pareto-optimal set are employed to design the NSGA II based ParaOptiMUSIG activation function. The optimized class boundaries for individual colour component in the selected chromosomes are applied to generate the individual OptiMUSIG activation function for that colour component, viz., the class boundaries for the red component is employed to generate OptiMUSIG activation function for red and so on. The NSGA II based ParaOptiMUSIG activation function is rendered using Eq. 6.1.

6.3.1.3 Input of True Colour Image Pixel Values to the Source Layer of the PSONN Architecture

The source layer of the PSONN architecture is inputted with the pixel intensity levels of the true colour image. The input pixel intensities of the true colour image are assigned to each of the neurons of the source layer.

6.3.1.4 Distribution of the Colour Component Images to Three Individual SONNs

The pixel intensity levels of the input colour image are segregated into three individual primary colour components and the three individual three-layer component SONNs are fed with these independent primary colour components, viz. the red component is utilized to one SONN, the green component to another SONN and the remaining SONN accepts the blue component information at their respective input layers. The fixed interconnections of the respective SONNs with the source layer are responsible for this scenario.

6.3.1.5 Segmentation of Colour Component Images by Individual SONNs

The corresponding SONN architecture guided by the designed ParaOptiMUSIG activation function at the constituent primitives/neurons is applied to segment the individual colour component of the true colour images. Depending on the number of transition lobes of the ParaOptiMUSIG activation function, the neurons of the different layers of individual three-layer SONN architecture render different output colour component level responses. The subnormal linear index of fuzziness [72] is employed to decide the system errors at the corresponding output layers as the network has no *a priori* knowledge about the outputs. Like the grayscale image segmentation, these errors are employed to adjust the interconnection weights between different layers using the standard backpropagation algorithm [37, 47]. After attaining stabilization in the corresponding networks, the respective output layers of the independent SONNs generate the final colour component images.

6.3.1.6 Fusion of Individual Segmented Component Outputs into a True Colour Image at the Sink Layer of the PSONN Architecture

In this final stage, the segmented outputs deduced at the three output layers of the three independent three-layer SONN architectures are fused at the sink layer of the PSONN architecture to derive the segmented true colour image. The number of segments is a combination of the number of transition lobes of the designed NSGA II based ParaOptiMUSIG activation functions used during component level segmentation.

6.3.2 Result Analysis

The multilevel true colour image segmentation approach using the NSGA II based ParaOptiMUSIG activation function with the help of PSONN architecture is demonstrated using the colour version of two real-life images, viz., Baboon and Peppers each of dimension 128×128 and a colour image of a human brain using a magnetic resonance imaging (MRI) machine [36] of dimension 170×170. Experiments have been carried with $K = \{6, 8\}$ classes. Both the activation functions have been designed with a fixed slope, $\lambda = \{2, 4\}$. But the results are reported for 8 classes and a fixed slope of $\lambda = 4$. The segmentation effectivity of this approach is demonstrated in the following subsection and it is compared with the segmentation derived by means of the conventional MUSIG activation function with same number of class responses and heuristic class levels. The quantitative comparison of the above-mentioned methods are manifested in the next section and in the next to next section, the comparative study as regards to its efficacy in the segmentation of test images of the segmented outputs by the narrated methods and the conventional MUSIG activation function based methods are detailed.

6.3.2.1 Quantitative Performance Analysis of Segmentation

In this section, the efficiency of the NSGA II based ParaOptiMUSIG activation function and the conventional MUSIG activation functions for $K = 8$ has been exemplified quantitatively. Three segmentation efficiency functions viz. correlation coefficient (ρ), and two empirical evaluation functions (F and F') have been employed in the NSGA II based approach to generate the NSGA II based ParaOptiMUSIG activation functions and another empirical evaluation function (Q) has been employed for the selection of the chromosome applied in the corresponding network in this method. The results obtained with the NSGA II based ParaOptiMUSIG activation function have been discussed in the next subsection and after that the corresponding results obtained with the conventional fixed class response based MUSIG activation function are furnished.

1. *Segmentation evaluation using NSGA II based ParaOptiMUSIG activation function for colour images*: The optimized set of class boundaries for the colour images obtained using NSGA II based algorithm with the three evaluation functions (ρ, F and F') are depicted in Tables 6.9, 6.10 and 6.11. Three fitness values (ρ, F and F') of individual chromosome set are also reported in these tables. Two set of Pareto-optimal set of chromosomes of each images are tabulated in the above mentioned tables. The Q values of individual set of class boundaries for colour images have been used for the gradation system for that particular set of class boundaries are presented in Tables 6.12, 6.13 and 6.14. The last columns of the Tables 6.9, 6.10 and 6.11 show the quality measures κ [graded on a scale of 1 (best) onwards] obtained by the segmentation of the test images based on the corresponding set of optimized class boundaries. The first four good Q valued chromosomes are indicated in those tables for easy reckoning. In this article, all evaluation function values are tabulated in normalized form.

Table 6.9 NSGA II based set of optimized class boundaries and corresponding evaluated segmentation quality measures for 8 classes of colour Baboon image

Set no		Class level	ρ	F	F'	κ
1	(i)	R = {27, 95, 104, 137, 173, 197, 239, 252}; G = {24, 76, 93, 119, 133, 158, 182, 209}; B = {0, 45, 72, 97, 127, 161, 221, 249}	0.9820	0.2016	0.0713	
	(ii)	R = {27, 95, 100, 139, 148, 186, 219, 252}; G = {24, 78, 91, 119, 135, 165, 182, 209}; B = {0, 45, 72, 97, 127, 161, 221, 249}	0.9821	0.2018	0.0714	2
	(iii)	R = {27, 95, 97, 135, 148, 184, 217, 252}; G = {24, 76, 93, 119, 135, 163, 182, 209}; B = {0, 45, 78, 109, 136, 161, 221, 249}	0.9856	0.2158	0.0763	
	(iv)	R = {27, 95, 96, 137, 148, 186, 219, 252}; G = {24, 76, 93, 119, 135, 165, 182, 209}; B = {0, 45, 72, 97, 127, 161, 221, 249}	0.9856	0.2040	0.0721	1
	(v)	R = {27, 99, 104, 137, 173, 186, 219, 252}; G = {24, 76, 91, 119, 135, 165, 182, 209}; B = {0, 45, 72, 97, 127, 161, 221, 249}	0.9815	0.2016	0.0713	
	(vi)	R = {27, 99, 104, 135, 173, 197, 239, 252}; G = {24, 76, 91, 119, 133, 158, 182, 209}; B = {0, 45, 72, 97, 127, 161, 221, 249}	0.9847	0.2027	0.0717	
	(vii)	R = {27, 99, 106, 137, 148, 186, 219, 252}; G = {24, 78, 91, 119, 135, 165, 182, 209}; B = {0, 45, 72, 97, 127, 161, 221, 249}	0.9856	0.2035	0.0720	
	(viii)	R = {27, 95, 97, 135, 148, 200, 208, 252}; G = {24, 76, 93, 119, 135, 145, 176, 209}; B = {0, 45, 78, 109, 138, 175, 208, 249}	0.9856	0.2090	0.0739	
	(ix)	R = {27, 95, 98, 137, 148, 198, 210, 252}; G = {24, 76, 93, 115, 131, 145, 176, 209}; B = {0, 45, 80, 107, 138, 175, 208, 249}	0.9855	0.2033	0.0719	4
	(x)	R = {27, 95, 98, 139, 148, 186, 217, 252}; G = {24, 76, 91, 119, 137, 163, 182, 209}; B = {0, 45, 72, 97, 125, 163, 221, 249}	0.9820	0.2018	0.0713	3
2	(i)	R = {27, 64, 94, 124, 140, 180, 192, 252}; G = {24, 74, 99, 123, 145, 172, 190, 209}; B = {0, 49, 75, 102, 132, 165, 218, 249}	0.9856	0.2050	0.0725	
	(ii)	R = {27, 64, 94, 126, 140, 182, 192, 252}; G = {24, 74, 97, 123, 141, 170, 190, 209}; B = {0, 49, 75, 102, 132, 163, 222, 249}	0.9856	0.2055	0.0726	3
	(iii)	R = {27, 66, 97, 124, 140, 180, 192, 252}; G = {24, 74, 91, 123, 141, 170, 190, 209}; B = {0, 47, 75, 102, 132, 165, 216, 249}	0.9855	0.2043	0.0722	3

(continued)

Table 6.9 (continued)

Set no	Class level	ρ	F	F'	κ
(iv)	R = {27, 64, 94, 126, 140, 180, 190, 252}; G = {24, 74, 97, 123, 141, 172, 190, 209}; B = {0, 49, 75, 102, 132, 163, 218, 249}	0.9856	0.2047	0.0724	
(v)	R = {27, 66, 97, 124, 140, 180, 192, 252}; G = {24, 76, 91, 123, 141, 170, 190, 209}; B = {0, 47, 77, 102, 132, 165, 216, 249}	0.9855	0.2045	0.0723	1
(vi)	R = {27, 66, 97, 124, 140, 180, 190, 252}; G = {24, 74, 99, 121, 141, 172, 190, 209}; B = {0, 47, 75, 102, 132, 163, 218, 249}	0.9855	0.2046	0.0723	
(vii)	R = {27, 66, 97, 124, 140, 180, 192, 252}; G = {24, 74, 97, 123, 141, 170, 190, 209}; B = {0, 47, 75, 102, 132, 165, 216, 249}	0.9856	0.2052	0.0725	2
(viii)	R = {27, 66, 97, 124, 140, 182, 192, 252}; G = {24, 74, 97, 121, 141, 170, 190, 209}; B = {0, 47, 75, 102, 132, 163, 222, 249}	0.9856	0.2052	0.0726	
(ix)	R = {27, 66, 97, 124, 140, 180, 190, 252}; G = {24, 76, 99, 123, 141, 172, 190, 209}; B = {0, 47, 75, 102, 132, 163, 218, 249}	0.9854	0.2039	0.0721	
(x)	R = {27, 66, 97, 124, 142, 180, 190, 252}; G = {24, 74, 97, 123, 139, 172, 190, 209}; B = {0, 47, 75, 102, 130, 163, 220, 249}	0.9855	0.2043	0.0723	4
(xi)	R = {27, 64, 94, 124, 142, 180, 190, 252}; G = {24, 74, 99, 123, 145, 172, 190, 209}; B = {0, 49, 75, 102, 132, 163, 218, 249}	0.9856	0.2051	0.0725	
(xii)	R = {27, 64, 94, 124, 142, 180, 190, 252}; G = {24, 74, 97, 123, 141, 172, 190, 209}; B = {0, 47, 75, 102, 132, 163, 218, 249}	0.9855	0.2045	0.0723	

Table 6.10 NSGA II based set of optimized class boundaries and corresponding evaluated segmentation quality measures for 8 classes of colour Peppers image

Set no	Class level		ρ	F	F'	κ
1	(i)	R = {13, 66, 83, 112, 119, 151, 161, 224}; G = {0, 35, 49, 99, 121, 165, 167, 234}; B = {0, 37, 60, 84, 113, 159, 194, 224}	0.9799	0.6321	0.2235	1
	(ii)	R = {13, 66, 83, 112, 143, 156, 186, 224}; G = {0, 35, 51, 106, 129, 174, 191, 234}; B = {0, 37, 60, 82, 101, 132, 167, 224}	0.9815	0.6531	0.2309	
	(iii)	R = {13, 66, 81, 112, 119, 151, 161, 224}; G = {0, 37, 51, 108, 121, 163, 167, 234}; B = {0, 37, 60, 84, 113, 157, 194, 224}	0.9801	0.6336	0.2240	
	(iv)	R = {13, 66, 81, 114, 143, 156, 188, 224}; G = {0, 35, 49, 106, 129, 176, 193, 234}; B = {0, 37, 60, 82, 101, 134, 169, 224}	0.9815	0.6527	0.2308	
	(v)	R = {13, 68, 81, 112, 143, 156, 188, 224}; G = {0, 35, 51, 106, 129, 176, 193, 234}; B = {0, 35, 58, 80, 101, 132, 167, 224}	0.9816	0.6539	0.2312	
	(vi)	R = {13, 66, 83, 114, 149, 162, 192, 224}; G = {0, 35, 51, 108, 133, 180, 193, 234}; B = {0, 37, 60, 82, 103, 136, 167, 224}	0.9816	0.6570	0.2323	
	(vii)	R = {13, 66, 83, 112, 143, 156, 188, 224}; G = {0, 35, 51, 106, 129, 176, 193, 234}; B = {0, 35, 58, 80, 101, 134, 169, 224}	0.9814	0.6500	0.2298	
	(viii)	R = {13, 66, 83, 112, 119, 147, 159, 224}; G = {0, 35, 51, 106, 121, 165, 167, 234}; B = {0, 37, 60, 84, 113, 153, 194, 224}	0.9816	0.6580	0.2326	
	(ix)	R = {13, 66, 83, 112, 145, 160, 188, 224}; G = {0, 37, 51, 106, 129, 176, 193, 234}; B = {0, 37, 60, 82, 103, 136, 165, 224};	0.9802	0.6381	0.2256	
	(x)	R = {13, 66, 83, 112, 145, 154, 190, 224}; G = {0, 35, 53, 104, 129, 176, 193, 234}; B = {0, 35, 58, 80, 101, 132, 169, 224}	0.9816	0.6660	0.2355	
	(xi)	R = {13, 66, 83, 112, 119, 151, 161, 224}; G = {0, 37, 51, 108, 121, 163, 167, 234}; B = {0, 37, 60, 84, 113, 157, 194, 224}	0.9812	0.6480	0.2291	2
	(xii)	R = {13, 68, 81, 110, 145, 156, 188, 224}; G = {0, 35, 51, 106, 127, 176, 193, 234}; B = {0, 35, 58, 80, 101, 132, 167, 224}	0.9816	0.6618	0.2340	
	(xiii)	R = {13, 66, 79, 114, 149, 162, 192, 224}; G = {0, 35, 49, 106, 133, 180, 193, 234}; B = {0, 37, 60, 82, 103, 134, 167, 224}	0.9801	0.6346	0.2243	

(continued)

Table 6.10 (continued)

Set no	Class level	ρ	F	F'	κ
	(xiv)	R = {13, 66, 85, 112, 119, 151, 161, 224}; G = {0, 41, 49, 99, 121, 163, 167, 234}; B = {0, 37, 62, 84, 113, 157, 194, 224}	0.9816	0.6570	0.2323	
	(xv)	R = {13, 66, 81, 114, 143, 160, 190, 224}; G = {0, 35, 51, 106, 133, 180, 193, 234}; B = {0, 37, 60, 82, 103, 136, 167, 224}	0.9816	0.6577	0.2325	
	(xvi)	R = {13, 66, 81, 112, 119, 151, 161, 224}; G = {0, 41, 51, 99, 121, 165, 169, 234}; B = {0, 37, 60, 87, 119, 157, 194, 224}	0.9814	0.6520	0.2305	4
	(xvii)	R = {13, 66, 83, 112, 119, 151, 161, 224}; G = {0, 35, 51, 108, 121, 163, 167, 234}; B = {0, 37, 60, 84, 113, 157, 194, 224}	0.9815	0.6528	0.2308	3
	(xviii)	R = {13, 64, 81, 112, 143, 156, 188, 224}; G = {0, 35, 51, 106, 129, 176, 193, 234}; B = {0, 35, 58, 80, 101, 134, 169, 224}	0.9800	0.6328	0.2237	
	(xix)	R = {13, 68, 81, 112, 143, 154, 190, 224}; G = {0, 35, 51, 104, 129, 176, 193, 234}; B = {0, 35, 58, 80, 101, 132, 169, 224}	0.9797	0.6286	0.2223	
	(xx)	R = {13, 66, 81, 114, 143, 156, 188, 224}; G = {0, 35, 49, 108, 129, 176, 193, 234}; B = {0, 37, 60, 80, 101, 134, 171, 224}	0.9815	0.6528	0.2308	
	(xxi)	R = {13, 66, 81, 114, 119, 151, 159, 224}; G = {0, 39, 47, 99, 121, 165, 167, 234}; B = {0, 37, 60, 86, 119, 157, 194, 224}	0.9814	0.6504	0.2299	
	(xxii)	R = {13, 66, 83, 112, 149, 154, 190, 224}; G = {0, 35, 53, 92, 129, 174, 193, 234}; B = {0, 37, 60, 80, 101, 130, 169, 224}	0.9815	0.6530	0.2309	
	(xxiii)	R = {13, 66, 81, 112, 149, 156, 190, 224}; G = {0, 41, 51, 99, 129, 176, 193, 234}; B = {0, 35, 58, 80, 101, 130, 169, 224}	0.9797	0.6292	0.2225	
	(xxiv)	R = {13, 66, 83, 112, 143, 156, 186, 224}; G = {0, 37, 51, 106, 129, 174, 191, 234}; B = {0, 37, 60, 82, 101, 132, 167, 224}	0.9801	0.6352	0.2246	
2	(i)	R = {13, 73, 82, 123, 135, 173, 187, 224}; G = {0, 21, 54, 91, 113, 157, 166, 234}; B = {0, 37, 60, 83, 110, 145, 179, 224}	0.9815	0.6255	0.2211	
	(ii)	R = {13, 73, 78, 123, 135, 173, 185, 224}; G = {0, 21, 54, 91, 111, 155, 166, 234}; B = {0, 37, 60, 83, 108, 143, 179, 224}	0.9816	0.6578	0.2326	
	(iii)	R = {13, 73, 78, 123, 135, 173, 187, 224}; G = {0, 21, 54, 91, 113, 157, 166, 234}; B = {0, 37, 60, 83, 110, 145, 179, 224}	0.9812	0.6200	0.2192	2

(continued)

Table 6.10 (continued)

Set no	Class level	ρ	F	F'	κ
(iv)	R = {13, 73, 80, 125, 137, 176, 185, 224}; G = {0, 19, 58, 91, 142, 162, 209, 234}; B = {0, 37, 60, 83, 104, 139, 174, 224}	0.9815	0.6202	0.2193	1
(v)	R = {13, 73, 80, 123, 135, 176, 187, 224}; G = {0, 19, 58, 93, 144, 162, 209, 234}; B = {0, 37, 60, 81, 104, 139, 174, 224}	0.9811	0.6191	0.2189	
(vi)	R = {13, 73, 78, 123, 133, 173, 187, 224}; G = {0, 21, 54, 91, 111, 155, 166, 234}; B = {0, 37, 60, 83, 106, 145, 179, 224}	0.9812	0.6192	0.2189	
(vii)	R = {13, 65, 69, 125, 133, 173, 187, 224}; G = {0, 35, 50, 91, 113, 157, 166, 234}; B = {0, 38, 60, 83, 106, 145, 179, 224}	0.9817	0.6585	0.2328	
(viii)	R = {13, 65, 80, 123, 137, 173, 187, 224}; G = {0, 35, 50, 93, 113, 157, 166, 234}; B = {0, 38, 60, 83, 104, 145, 179, 224}	0.9817	0.6603	0.2335	
(ix)	R = {13, 73, 78, 123, 135, 173, 187, 224}; G = {0, 21, 54, 91, 111, 157, 168, 234}; B = {0, 37, 60, 83, 108, 147, 179, 224}	0.9815	0.6212	0.2196	
(x)	R = {13, 65, 67, 125, 133, 173, 187, 224}; G = {0, 33, 50, 91, 113, 155, 166, 234}; B = {0, 38, 60, 83, 106, 143, 179, 224}	0.9817	0.6591	0.2330	
(xi)	R = {13, 73, 80, 123, 135, 173, 187, 224}; G = {0, 21, 56, 91, 113, 157, 166, 234}; B = {0, 37, 60, 83, 110, 145, 179, 224}	0.9811	0.6148	0.2174	
(xii)	R = {13, 71, 82, 125, 137, 176, 185, 224}; G = {0, 23, 58, 89, 142, 162, 209, 234}; B = {0, 37, 60, 83, 102, 139, 174, 224}	0.9810	0.6134	0.2169	
(xiii)	R = {13, 65, 69, 123, 137, 173, 187, 224}; G = {0, 35, 50, 93, 113, 157, 166, 234}; B = {0, 38, 60, 83, 106, 145, 179, 224}	0.9813	0.6202	0.2193	3
(xiv)	R = {13, 65, 80, 125, 133, 175, 187, 224}; G = {0, 35, 50, 91, 113, 159, 166, 234}; B = {0, 38, 60, 83, 104, 147, 179, 224}	0.9814	0.6204	0.2194	
(xv)	R = {13, 73, 82, 123, 133, 173, 187, 224}; G = {0, 21, 58, 91, 113, 155, 166, 234}; B = {0, 37, 60, 83, 106, 145, 179, 224}	0.9817	0.6587	0.2329	
(xvi)	R = {13, 73, 82, 125, 137, 176, 185, 224}; G = {0, 19, 58, 91, 142, 162, 209, 234}; B = {0, 37, 60, 83, 104, 139, 174, 224}	0.9817	0.6607	0.2336	
(xvii)	R = {13, 73, 82, 123, 137, 171, 187, 224}; G = {0, 19, 56, 91, 142, 157, 166, 234}; B = {0, 37, 60, 83, 104, 143, 179, 224}	0.9814	0.6238	0.2205	4

Table 6.11 NSGA II based set of optimized class boundaries and corresponding evaluated segmentation quality measures for 8 classes of colour Brain MRI image

Set no		Class level	ρ	F	F'	κ
1	(i)	R = {2, 31, 71, 119, 135, 193, 213, 254}; G = {3, 43, 69, 112, 143, 182, 208, 236}; B = {0, 17, 52, 92, 128, 166, 206, 254}	0.9926	0.7877	0.6538	1
	(ii)	R = {2, 31, 75, 117, 135, 191, 213, 254}; G = {3, 45, 71, 116, 143, 180, 208, 236}; B = {0, 17, 52, 92, 126, 166, 206, 254}	0.9926	0.8027	0.6764	4
	(iii)	R = {2, 55, 63, 121, 137, 191, 213, 254}; G = {3, 44, 71, 112, 143, 180, 208, 236}; B= {0, 18, 52, 92, 128, 166, 206, 254}	0.9923	0.8454	0.8706	
	(iv)	R = {2, 55, 63, 121, 135, 191, 211, 254}; G = {3, 44, 71, 112, 143, 178, 208, 236}; B= {0, 18, 52, 92, 126, 166, 206, 254}	0.9915	0.9393	0.9430	
	(v)	R = {2, 55, 63, 123, 135, 191, 211, 254}; G = {3, 44, 73, 114, 143, 178, 208, 236}; B = {0, 18, 54, 93, 126, 166, 206, 254}	0.9906	0.9520	1.0000	
	(vi)	R = {2, 31, 73, 115, 135, 191, 213, 254}; G = {3, 43, 71, 116, 143, 180, 208, 236}; B = {0, 17, 52, 92, 126, 166, 206, 254}	0.9908	0.8191	0.6978	3
	(vii)	R = {2, 31, 71, 115, 135, 191, 213, 254}; G = {3, 43, 69, 116, 141, 180, 208, 236}; B = {0, 17, 48, 88, 126, 166, 206, 254}	0.9908	0.8265	0.6878	2
2	(i)	R = {2, 39, 63, 112, 138, 182, 196, 254}; G = {3, 46, 61, 113, 122, 169, 179, 236}; B = {0, 17, 52, 94, 131, 171, 213, 254}	0.9926	0.8069	0.6901	
	(ii)	R = {2, 39, 63, 110, 138, 182, 198, 254}; G = {3, 46, 61, 113, 122, 173, 179, 236}; B = {0, 17, 56, 94, 131, 175, 213, 254}	0.9924	0.7480	0.6455	2
	(iii)	R = {2, 41, 59, 112, 138, 182, 194, 254}; G = {3, 46, 59, 111, 120, 173, 179, 236}; B = {0, 17, 48, 92, 131, 171, 213, 254}	0.9915	0.8574	0.7681	
	(iv)	R = {2, 39, 63, 104, 134, 182, 194, 254}; G = {3, 46, 59, 113, 120, 173, 179, 236}; B = {0, 17, 52, 94, 131, 171, 213, 254}	0.9912	0.8002	0.7066	

(continued)

Table 6.11 (continued)

Set no		Class level	ρ	F	F'	κ
(v)		R = {2, 39, 63, 112, 132, 182, 198, 254}; G = {3, 17, 48, 92, 131, 171, 213, 236}; B = {0, 17, 56, 94, 131, 175, 213, 254}	0.9918	0.7847	0.6252	1
(vi)		R = {2, 41, 59, 104, 134, 182, 194, 254}; G = {3, 46, 59, 113, 120, 173, 179, 236}; B = {0, 17, 48, 92, 131, 171, 213, 254}	0.9918	0.8384	0.7312	
(vii)		R = {2, 39, 61, 112, 132, 182, 198, 254}; G = {3, 46, 61, 113, 122, 173, 177, 236}; B = {0, 17, 52, 94, 131, 175, 213, 254}	0.9915	0.8001	0.7216	4
(viii)		R = {2, 39, 65, 112, 132, 182, 198, 254}; G = {3, 46, 59, 113, 124, 173, 177, 236}; B = {0, 17, 56, 94, 131, 175, 213, 254}	0.9919	0.7854	0.6624	3
(ix)		R = {2, 39, 63, 112, 140, 182, 198, 254}; G = {3, 46, 61, 111, 122, 173, 177, 236}; B = {0, 17, 52, 94, 131, 173, 213, 254}	0.9921	0.8088	0.7703	

Table 6.12 The evaluated segmentation quality measure Q of the each optimized class boundaries of Table 6.9

Set	No	(i)	(ii)	(iii)	(iv)	(v)	(vi)	(vii)
1	Q	0.3996	0.2863	0.4020	0.2844	0.4182	0.4008	0.3867
	No	(viii)	(ix)	(x)				
	Q	0.2949	0.2929	0.2875				
2	No	(i)	(ii)	(iii)	(iv)	(v)	(vi)	(vii)
	Q	0.3225	0.3262	0.3171	0.3291	0.3144	0.3206	0.3162
	No	(viii)	(ix)	(x)	(xi)	(xii)		
	Q	0.3237	0.3229	0.3195	0.3225	0.3262		

Table 6.13 The evaluated segmentation quality measure Q of the each optimized class boundaries of Table 6.10

Set	No	(i)	(ii)	(iii)	(iv)	(v)	(vi)	(vii)
1	Q	0.6371	0.6868	0.6470	0.7013	0.6721	0.6857	0.6739
	No	(viii)	(ix)	(x)	(xi)	(xii)	(xiii)	(xiv)
	Q	0.6661	0.6680	0.6672	0.6429	0.6709	0.6873	0.6501
	No	(xv)	(xvi)	(xvii)	(xviii)	(xix)	(xx)	(xxi)
	Q	0.6624	0.6457	0.6446	0.6687	0.6785	0.6924	0.6757
	No	(xxii)	(xxiii)	(xxiv)				
	Q	0.6702	0.6770	0.6775				
2	No	(i)	(ii)	(iii)	(iv)	(v)	(vi)	(vii)
	Q	0.6456	0.6815	0.6882	0.6807	0.7001	0.6799	0.6994
	No	(viii)	(ix)	(x)	(xi)	(xii)	(xiii)	(xiv)
	Q	0.6391	0.7059	0.6986	0.6589	0.6465	0.6788	0.6456
	No	(xv)	(xvi)	(xvii)				
	Q	0.6815	0.6882	0.6807				

Table 6.14 The evaluated segmentation quality measure Q of the each optimized class boundaries of Table 6.11

Set	No	(i)	(ii)	(iii)	(iv)	(v)	(vi)	(vii)
1	Q	0.8230	0.8477	0.8769	0.8856	0.8902	0.8382	0.8338
2	No	(i)	(ii)	(iii)	(iv)	(v)	(vi)	(vii)
	Q	0.9025	0.8893	0.9072	0.9001	0.8889	0.9028	0.8934
	No	(viii)	(ix)					
	Q	0.8905	0.9005					

Table 6.15 Fixed class boundaries and corresponding evaluated segmentation quality measures for 8 classes of colour Baboon and Peppers image

Image	Set no	Class level	ρ	F	F'
Baboon	1	R = {27, 60, 110, 130, 160, 190, 210, 252}; G = {24, 50, 70, 90, 115, 145, 175, 209}; B = {0, 45, 90, 116, 145, 175, 200, 249}	0.9422	0.7881	0.7881
	2	R = {27, 40, 60, 110, 180, 215, 225, 252}; G = {24, 50, 70, 100, 145, 175, 190, 209}; B = {0, 35, 90, 160, 185, 205, 230, 249}	0.9297	1.0000	1.0000
	3	R = {27, 45, 75, 130, 150, 200, 235, 252}; G = {24, 55, 70, 85, 130, 170, 185, 209}; B = {0, 55, 80, 115, 125, 185, 225, 249}	0.9406	0.7408	0.7408
	4	R = {27, 50, 70, 120, 185, 215, 230, 252}; G = {24, 45, 65, 95, 140, 175, 195, 209}; B = {0, 35, 100, 150, 185, 215, 230, 249}	0.9314	0.9800	0.9800
Peppers	1	R = {0, 20, 60, 120, 150, 190, 210, 253}; G = {0, 40, 75, 90, 190, 215, 230, 255}; B = {0, 35, 60, 100, 125, 160, 175, 245}	0.9745	1.0000	1.0000
	2	R = {0, 25, 70, 130, 145, 175, 205, 253}; G = {0, 45, 80, 100, 170, 200, 225, 255}; B = {0, 30, 60, 90, 115, 150, 180, 245}	0.9803	0.8734	0.8734
	3	R = {0, 30, 75, 135, 150, 180, 215, 253}; G = {0, 55, 90, 105, 175, 195, 215, 255}; B = {0, 40, 65, 105, 135, 170, 195, 245}	0.9761	0.8612	0.8612
	4	R = {0, 50, 75, 125, 155, 192, 220, 253}; G = {0, 60, 85, 130, 180, 206, 235, 255}; B = {0, 37, 60, 85, 109, 150, 195, 245}	0.9847	0.8430	0.8430
MRI	1	R = {2, 30, 60, 100, 135, 195, 215, 254}; G = {3, 30, 60, 100, 135, 195, 215, 236}; B = {0, 30, 60, 100, 135, 195, 215, 254}	0.9887	0.7895	0.5849
	2	R = {2, 35, 70, 95, 130, 190, 218, 254}; G = {3, 35, 70, 95, 130, 190, 218, 236}; B = {0, 35, 70, 95, 130, 190, 218, 254}	0.9867	1.0000	0.8805
	3	R = {2, 25, 65, 110, 145, 200, 220, 254}; G = {25, 65, 110, 145, 200, 220, 236}; B = {0, 25, 65, 110, 145, 200, 220, 254}	0.9902	0.6877	0.4739
	4	R = {2, 28, 55, 105, 140, 210, 225, 254}; G = {3, 28, 55, 105, 140, 210, 225, 236}; B = {0, 28, 55, 105, 140, 210, 225, 254}	0.9868	0.9102	0.9032

Table 6.16 The evaluated segmentation quality measure Q of the each fixed class boundaries of Table 6.15

Image	Set no	Q	Image	Set no	Q	Image	Set no	Q
Baboon	1	0.3917	Peppers	1	0.8744	MRI	1	0.9054
	2	0.8139		2	0.8978		2	1.0000
	3	1.0000		3	0.8995		3	0.7627
	4	0.4959		4	1.0000		4	0.8158

2. *Segmentation evaluation using MUSIG activation function for colour images*: The heuristically selected class boundaries are applied in the conventional MUSIG activation function for colour images and they are depicted in Table 6.15. The evaluation procedure is exercised for the individual set of class boundaries along with the same evaluation functions and the results are reported in those tables. After the segmentation process using the MUSIG activation function, the quality of the segmented images are evaluated using the Q evaluation function and the results are tabulated in Table 6.16 for colour images.

It is clearly revealed from Tables 6.12, 6.13, 6.14 and 6.16 that the fitness values derived by the NSGA II based ParaOptiMUSIG activation function for colour images are better than those obtained using the conventional MUSIG activation function. In most of the cases, the Q value in Tables 6.12, 6.13 and 6.14 is lower than the same value in Table 6.16 for both images. From Chap. 3 discussion, it is to be noted that the lower value of Q signifies better segmentation. It is evident from those tables that the image segmentation done by the illustrated method generates lower Q value than the image segmentation done by the heuristically selected class levels. In another way, the ρ, F and F' values of Tables 6.9, 6.10 and 6.11 derived by the NSGA II based ParaOptiMUSIG activation function for colour images are better than those values, reported in Table 6.15 by the conventional MUSIG activation function for the respective images.

6.3.3 Image Segmentation Outputs

The segmented colour output images obtained for the $K = 8$ classes, with the optimized approach vis-a-vis those prevailed with the heuristically chosen class boundaries, are demonstrated in this section.

6.3.3.1 Segmented Outputs Using Optimized Approach

The first four better results on the basis of the evaluation metric, Q, value are applied in the network to segment the test images. The segmented colour test images derived with the PSONN architecture using the NSGA II based ParaOptiMUSIG activation function for $K = 8$ are shown in Figs. 6.8, 6.9 and 6.10.

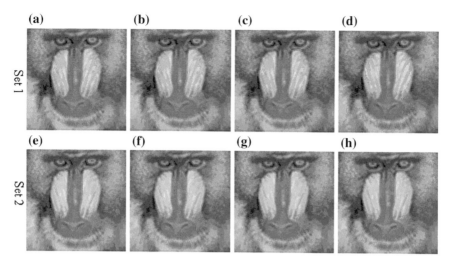

Fig. 6.8 8-class segmented 128 × 128 colour Baboon image with the optimized class levels referring to **a–d** set 1 **e–h** set 2 of Table 6.9 for first four better quality measure Q with NSGA II based ParaOptiMUSIG activation function

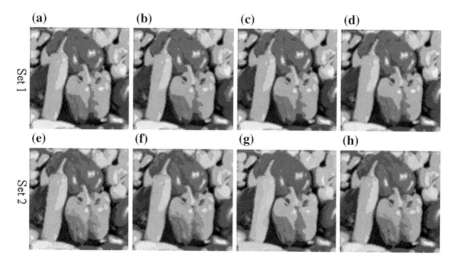

Fig. 6.9 8-class segmented 128 × 128 colour Peppers image with the optimized class levels referring to **a–d** set 1 **e–h** set 2 of Table 6.10 for first four better quality measure Q with NSGA II based ParaOptiMUSIG activation function

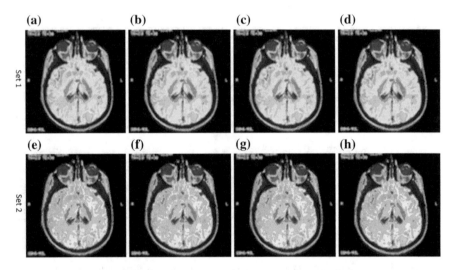

Fig. 6.10 8-class segmented 170×170 colour Brain MRI image with the optimized class levels referring to **a–d** set 1 **e–h** set 2 of Table 6.11 for first four better quality measure Q with NSGA II based ParaOptiMUSIG activation function

6.3.3.2   Segmented Outputs Using MUSIG Activation Function

The fixed class responses are applied to generate the segmented colour test images obtained with the PSONN architecture characterised by the conventional MUSIG activation function. The segmented output colour images for $K = 8$ are shown in Figs. 6.11 and 6.12.

From the results obtained, the image segmentation method using the conventional MUSIG activation function is outperformed by the image segmentation method by the NSGA II based ParaOptiMUSIG activation function for colour images as regards to the segmentation quality of the images for the different number of classes. It is also evident that the present approach incorporates the image heterogeneity as it can handle a wide variety of image intensity distribution prevalent in real life.

6.4   Discussions and Conclusion

In this chapter, colour image segmentation is talked about with the help of the parallel self-organizing neural network (PSONN) architecture. It has been noted that the parallel optimized multilevel MUSIG (ParaOptiMUSIG) activation function incorporates the parallelism property in the PSONN architecture. It has been observed that a set of class boundaries may generate good segmentation in respect of a certain criterion though it may or may not guaranteed that the same set of class boundaries will give a good segmentation based on another criterion. In the first part of the chapter, a

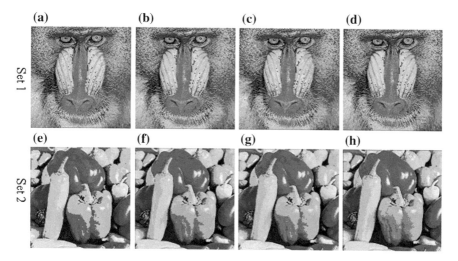

Fig. 6.11 8-class segmented 170×170 **a–d** colour Baboon image and **e–h** colour Peppers image with the fixed class levels of Table 6.16 with MUSIG activation function

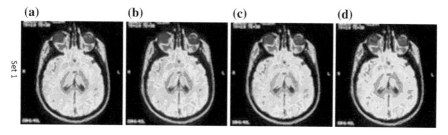

Fig. 6.12 8-class segmented 170×170 **a–d** colour Brain MRI image with the fixed class levels of Table 6.16 with MUSIG activation function

multiobjective genetic algorithm based ParaOptiMUSIG activation function is presented and a NSGA II based ParaOptiMUSIG activation function are purported in the second part of this chapter for true colour image segmentation. In both methods, the optimized class boundaries are derived on the basis of different image segmentation evaluation criterions. The selection of a particular set of class boundaries is performed based on another quantitative quality measure that is applied in the network. The performance of the multiobjective genetic algorithm based ParaOptiMUSIG activation function and the NSGA II based ParaOptiMUSIG activation function for the segmentation of real-life colour images show superior performance in most of the cases as compared to the conventional MUSIG activation function with heuristic class responses.

Up to this point, we started our segmentation method from a predefined number of classes but it may or may not generate good results. It may happen that we may get better result if we increase or decrease the number of classes and we are also unaware of the exact number of classes in those test images. It will be better that the segmentation process will start from a large number of classes and the process will find out the exact number of classes which is underneath in that test image. The next chapter attempts to work in that direction. The exact number of classes in a test image can be evaluated automatically by the algorithm and that algorithm is narrated and proved in the next chapter.

Chapter 7
Unsupervised Genetic Algorithm Based Automatic Image Segmentation and Data Clustering Technique Validated by Fuzzy Intercluster Hostility Index

7.1 Introduction

The main criterion of the segmentation methods discussed in the previous chapters, is to supply the number of classes beforehand. So, one can not be assured that number of classes generate the best segmented outputs. This chapter tries to overcome this drawback. This approach is quite effective to derive the exact number of classes from a large number of classes that will give the good segmented/cluster output.

Different clustering techniques are in vogue for evaluating the relationship between patterns by organizing patterns into clusters, so that the intracluster patterns are more similar among themselves than are the patterns belonging to different clusters, i.e. intercluster patterns. A good and extensive overview of clustering algorithms can be found in [87, 294, 295]. However, it is not an easy task to know the particular number of clusters in an unknown dataset [296]. The k-means algorithm is one of the simplest and most popular iterative clustering algorithm [87, 153, 297]. The procedure follows a simple and easy way to classify a given data set to a certain number of clusters. The cluster centroids are initialized randomly or derived from some a priori information. Each item in the data set is then assigned to the closest cluster. Finally, the centroids are reassigned according to the associated data point. This algorithm optimizes the distance criterion either by maximizing the intercluster separation or by minimizing the intracluster separation. The k-means algorithm is a data-dependent greedy algorithm that may converge to a suboptimal solution. An unsupervised clustering algorithm, using the k-means algorithm, is presented by Rosenberger et al. [154]. An improved k-means algorithm for clustering is presented in [155]. In this method, each data point is stored in a kd-tree and it is shown that the algorithm runs faster as the separation between the clusters increases. Pelleg and Moore [298] proposed another improved version of the k-means algorithm. Huang [299] proposed another partitioning algorithm that uses the simple matching coefficient measure to deal with categorical attributes. Recently, a generalization of the conventional k-means clustering algorithm has been presented in [300]. It

© Springer International Publishing AG 2016
S. De et al., *Hybrid Soft Computing for Multilevel Image and Data Segmentation*, Computational Intelligence Methods and Applications, DOI 10.1007/978-3-319-47524-0_7

is an unsupervised algorithm and is capable of clustering ellipse shaped as well as ball shaped data. Fuzzy c-means (FCM), originally introduced by Bezdek [199], is another most well-known data clustering technique. This is an iterative optimization algorithm that minimizes the sum of the Euclidean distances over all the clusters. Furthermore, the major limitations of FCM are that a priori specification of the number of clusters is required and that it often converges to suboptimal solutions, strongly depending on the choice of the initial values. Yang et al. [301] proposed a new clustering algorithm, referred to as the penalized FCM (PFCM) algorithm for application to images. Krishnapuram and Keller [4] proposed a clustering model named possibilistic c-means (PCM) to overcome the drawbacks of FCM. An algorithm based on fuzzy clustering to dynamically determine the number of clusters in a data set has been proposed by Lorette et al. [302]. It is a supervised method as it needs a parameter to be specified, which has a profound effect on the number of clusters to be generated. The Expectation-Maximization (EM) algorithm [303] is another soft clustering algorithm that assumes an underlying probability model, with parameters describing the probability of belongingness to a certain cluster. The drawbacks of this algorithm are that it is computationally intensive and the distributions of probabilities involve too many parameters. However, in most of the real life scenarios, the number of clusters in a data set is unknown. Most of these clustering algorithms are not capable of automatically and efficiently clustering various types of input datasets, especially when the number of clusters included in the data set tends to be large. Difficult problems such as these are referred to as unsupervised clustering or nonparametric clustering, and are often dealt with by employing an evolutionary approach.

Evolutionary algorithms, such as genetic algorithm [217], differential evolutionary algorithm [85], etc. are stochastic search heuristics inspired by Darwinian evolution and genetics. The clustering process can be optimized by using the algorithm (GA) in an evolutionary approach. Sheikh et al. [304] reported different types of GA based clustering algorithms. An application of GA for clustering has been presented by Raghavan and Birchand [305], which belongs to the first approach to minimize the squared error of a clustering. GA has been used to segment an image into different regions depending upon the piecewise linearity of the segments [221]. In this chapter, GA has been used to determine the hyperplanes as decision boundaries, which divide the dataset into a number of clusters. Srikanth et al. [306] proposed a supervised clustering algorithm by GA. Some classical clustering techniques are often hybridized with GA [307] to cluster unlabeled datasets. An improved genetic algorithm is presented by Katari et al. [308]. In this approach, GA is combined with the Nelder-Mead (NM) Simplex search and the k-means algorithm. Sarafis et al. [309] proposed a database clustering tool using rule-based genetic algorithm (RBCGA) that identifies clusters by a fixed set of clustering rules in the object attribute space. GAs are very useful for improving the performance of k-means algorithm. Krishna and Murty [5] combined GA with k-means and developed the genetic k-means algorithm (GKA) to find global optima. In this algorithm, the k-means algorithm is applied to determine the quality of GA candidate solutions. However, in most of the reported approaches, the number of clusters k is either set in advance or it is externally provided.

Nevertheless, the number of clusters from a previously unlabeled dataset can be found automatically. Automatic data clustering through determination of optimal number of clusters from the data content, is at the helm of affairs in the research community. Lack of knowledge regarding the underlying data distribution poses constraints in proper determination of the inherent number of clusters. Several algorithms for automatic clustering exist in the literature [310]. An automatic clustering is implemented by GA in biological data mining and information retrieval [311]. In this method, a cohesion-and-coupling metric is integrated with the proposed hybrid algorithm consisting of a genetic algorithm and a split-and-merge algorithm. Evolutionary strategy (ES) based approaches having the ability to cluster unlabeled datasets, is proposed by Lee and Antonsson [312]. In this approach, variable-length strings have been used to search for both the centroids and the optimal number of clusters. Different approaches, to dynamically cluster a dataset applying evolutionary programming, have been investigated in the recent past. Bandyopadhyay and Maulik [313] used a variable string-length genetic algorithm to determine to number of clusters automatically. The underlying string representation comprises both real numbers (for valid cluster centroids) and don't care symbols (for invalid ones).

Evolutionary computing techniques have been applied rigorously to determine clusters in complex data sets in the past few years. Differential Evolution (DE) algorithm is one such search and optimization algorithm which is more likely to find a function's true global optimum. The classical version of DE algorithm was proposed by Storn et al. [85]. In this approach, each component in the individual vector is represented by using real coding of floating point numbers. This technique is employed as a search procedure due to its fast convergence properties and ability to find the optimality of solutions. However, the application of algorithm to real-life situations is limited by the fact that the number of clusters must be provided externally. Das et al. [88] proposed an improved version of the classical form of DE algorithm by incorporating several modifications. The proposed automatic clustering differential evolution (ACDE) algorithm possesses the capability of clustering unlabeled datasets automatically. The active/inactive cluster centroids are selected based on a threshold value, which is associated with each and every cluster centroid. In this approach, a cluster centroid is designated as an active cluster if the corresponding threshold value is greater than 0.5. All other clusters with cluster centroids having thresholds less than 0.5 are made inactive, thereby arriving at an optimum active number of clusters. However, the basis for the choice of a threshold value of 0.5 is heuristic.

Several traditional approaches for determining the optimal number of clusters in a data set exist. The quality of partitioning for a range of cluster numbers is generally measured by a cluster validity index or a statistical-mathematical function. These are concerned with the determination of the optimal number of clusters followed by the investigation of the quality of the clustering results. The basic idea of determining the optimum number of classes is to employ several instances of the algorithm with varied number of classes as input and then to choose the partitioning of the dataset based on the best achievable validity measure. The maximum/minimum values of the validity measures suggest the appropriate partitions. Some well-known validity measures available in the literature are Davies-Bouldin's measure [281], Dunn's separation measure [314], Bezdek's partition coefficient [315], Xie-Beni's separation measure [316], Pakhira Bandyopadhyay Maulik (PBM) index [317], CS measure [318], etc. Optimization algorithms are mostly used to evaluate these cluster validity indices.

In this chapter, a novel GA based method for the automatic determination of the optimum number of clusters in images is presented. This approach is not only applied on different test images but also on different types of multidimensional data sets. In this process, valid/invalid clusters are selected from the test data context information, thereby obviating the need of thresholding mechanism. In the initial phase, the optimal number of clusters is determined from a given maximum number of clusters. Subsequently, the valid/invalid clusters are detected using a fuzzy intercluster hostility index [319–323], which is indicative of the degree of homogeneity/heterogeneity of the clusters. The optimal number of valid cluster centroids is obtained when the fuzzy intercluster hostility indices between the different clusters are lesser than the average fuzzy intercluster hostility index over all possible clusters [319–323]. The Davis-Bouldin cluster validity index [281] and CS measure [318] are used to compute the fitness of the individual chromosomes in this method. An application of this approach is demonstrated on three real life gray level images as well as three multidimensional data sets. The presented approach is compared with the ACDE algorithm [88].

7.2   Region Based Image Clustering

Image clustering is a process of segregating an image into different regions such that each region is uniform and homogeneous with respect to some features such as gray tone or texture. The adjacent regions of the clustered image are significantly different with respect to the characteristics on which they are uniform. Therefore, the result of a clustering is a labeled image, with each label corresponding to a region. Image clustering has to fulfill two constraints, i.e., the completeness of the segmentation and the connectivity of the regions. Haralick et al. [324] stressed upon the desirable characteristics that a good image segmentation should held with respect to gray level images. The unique object centric features to denote the different regions should be selected properly to cluster an image data. The resultant feature space acts as a knowledge base for segregating the image regions in terms of similarity/homogeneity. The properties of homogeneity and connectivity have to be maintained by each region in a clustered image. The pixels in a region are considered as homogeneous/heterogeneous if they satisfy a homogeneity/heterogeneity criterion in respect of the feature space.

Let M denote an image and let HM define a certain homogeneity predicate; then the segmentation of M is a partition of M into a set of N regions $RE_n = 1, 2, \ldots N$ such that [319–323]:

1. $\displaystyle\bigcup_{n=1}^{N} RE_i = M$ with $RE_i \cap RE_j \neq \phi, i \neq j$
2. $HM(RE_i) = true \; \forall \; RE_i$
3. $HM(RE_i \cup RE_j) = false$ for any two adjacent regions RE_i and RE_j.

This partitioning procedure has to cover the entire image and no given pixel can belong to two regions. Two adjacent regions are considered as homogeneous if the predicate is satisfied for the union of the two regions. Hence, these regions can be merged into a single region. On the other hand, a region is denoted as heterogeneous if the predicate is not satisfied and that region can be further split into sub-regions.

7.2.1 Fuzzy Intercluster Hostility Index

The image clustering characterizes the segregation of the constituent pixels based on uniformity/homogeneity of features. A cluster in a clustered image is denoted as homogeneous if the representative fuzzy membership values are close to each other. On the contrary, the sharp contrasting fuzzy membership values of the elements in the clusters denote that two regions are heterogeneous to each other. The fuzzy intercluster hostility index (ρ_{ij}) is applied to measure the intercluster heterogeneity between two regions/cluster (RE_i and RE_j) and it is denoted by [319–323]

$$\rho_{ij} = \triangle(RE_i, RE_j) = \frac{1}{|RE_i|.|RE_j|} \sum_{x \in RE_i} \sum_{y \in RE_j} |\mu(x) \otimes \mu(y)| \qquad (7.1)$$

where, $1 \leq (i, j) \leq K$, $\mu(x)$ and $\mu(y)$ are the intensity values of pixels x and y in the clusters/regions RE_i and RE_j, respectively. K denotes the maximum number of clusters. $\otimes$ denotes the difference operator on the representative fuzzy feature values. The cardinalities of the clusters/regions RE_i and RE_j are represented as $|RE_i|$ and $|RE_j|$, respectively. Hence [319–323],

- $\rho_{ij} = \triangle(RE_i, RE_i) = 0$, if RE_i contains homogeneous features.
- $\rho_{ij} = \triangle(RE_i, RE_j) = 1$, if RE_i and RE_j contains complementary features.
- $0 \leq \rho_{ij} \leq 1, \forall 1 \leq i, j \leq K$, in general.

A value of fuzzy intercluster hostility index close to unity means that the clusters are heterogeneous to each other. On the other hand, two clusters are referred as homogeneous to each other if the fuzzy intercluster hostility index between them is closer to zero.

7.3   Cluster Validity Indices

Clusters are detected by some clustering algorithms without having some a priori information and after that, some evaluation mechanism is required to measure the goodness of the final partition of the dataset [310, 325]. Basically, the quantitative evaluation of the results of the clustering algorithms is known as cluster validity indices and it is a statistical mathematical function. By using cluster validity indices, the number of clusters can be decided and the corresponding best partition can be derived. Traditionally, a clustering algorithm is iterated repeatedly to derive the optimum number of classes with a different number of classes as input and after that, the best partitioning of the data is selected on the basis of the best validity measure [88, 326]. The cluster validity is investigated by three approaches, viz. external criteria based approach, internal criteria based approach and relative criteria based approach [310, 327]. Two aspects of partitioning are kept in mind for determining the clustering evaluation and selection of an optimal clustering method [88, 310]:

1. Compactness: The characteristics of the data points within each cluster should be similar to each other as far as possible. A cluster's compactness can be measured by the fitness variance of the patterns.
2. Separation: The characteristics of the data points in different clusters are totally different. A cluster separation can be measured by the distance among the cluster centers.

Dunn's separation measure [314], Bezdek's partition coefficient [315], Xie-Beni's separation measure [316], Davies-Bouldin's measure [281], CS measure [318], Pakhira Bandyopadhyay Maulik (PBM) index [317] etc. are some of the renowned indices. The common characteristic of these indices is that the maximum or minimum values of these indices indicate the appropriate partitions. Different types of optimization algorithms such as GA, PSO, etc. apply these cluster validity indices for their optimizing character. In this chapter, only two validity measures are illustrated in detail and they have been applied in the automatic segmentation methods.

7.3.1   Davies-Bouldin (DB) Validity Index

DB index [281] is applied in a clustering procedure to decide which of the various features and channels would be of most use in the subsequent classification task. It is utilized as a validity criterion to evaluate compact and separate clusters. The ratio of within-cluster scatter to between-cluster separation is calculated by the DB-index. DB index uses both the clusters and their sample means. Let, $\mathcal{X} = \{X_1, \ldots, X_k\}$ be the data set and $U = \{U_1, \ldots, U_k\}$ be its K-class clustered output. The cluster similarity between the ith and jth clusters (R_{ij}), is evaluated along with the distance between the cluster centroids (D_{ij}). Cluster similarity is defined as [281]

$$R_{ij} = \frac{S_i + S_j}{D_{ij}} \tag{7.2}$$

where, S_i and S_j are the average dispersion of the ith and jth clusters, respectively. The intercluster distance is given by [281]

$$D_{ij} = \{\sum_{K=1}^{N} |m_{Ki} - m_{Kj}|^p\}^{\frac{1}{p}} \tag{7.3}$$

where, m_{Ki} is the centroid of cluster i. D_{ij} is the Minkowski distance between the centroids that characterize clusters i and j. The dispersion within a cluster is calculated from [281]

$$S_i = \{\frac{1}{T_i} \sum_{j}^{T_i} \|p_j - m_i\|_2^q\}^{\frac{1}{q}} \tag{7.4}$$

where, m_i is the centroid of cluster i. The centroid, m_i, is evaluated as $m_i = \frac{1}{T_i} \sum_{p \in C_i} p$ and T_i is the number of points in cluster C_i. Specifically, S_i, used in this article, is the standard deviation of the Euclidean distance between all data points in a cluster and the cluster's centroid. Finally, the DB-index is given by [281]

$$DB = \frac{1}{N} \sum_{i=1}^{N} R_i \text{ where } R_i = \max_{j \neq i} R_{ij} \tag{7.5}$$

A minimum DB-index implies proper and faithful clustering of a dataset.

7.3.2   CS Measure

CS measure [318], proposed by Chou et al., is applied to evaluate the validity of a clustered dataset. The cluster centroid m_i, is evaluated before utilizing in the CS measure by averaging the data vectors that belong to that cluster and it is denoted as [318]

$$\vec{m_i} = \frac{1}{U_i} \sum_{X_j \in U_i} \vec{X_j} \tag{7.6}$$

where U_i is the set of elements whose data points are allotted to the ith cluster and $|U_i|$ denotes the number of elements in U_i.

Then, the CS measure can be determined as [318]

$$
CS(K) = \frac{\frac{1}{K}\sum\limits_{i=1}^{K}[\frac{1}{U_i}\sum\limits_{\vec{X_j}\in U_i}\max\limits_{\vec{X_q}\in U_i}\{d(\vec{X_j},\vec{X_q})\}]}{\frac{1}{K}\sum\limits_{i=1}^{K}[\min\limits_{j\in k,j\neq i}\{d(\vec{m_i},\vec{m_j})\}]} = \frac{\sum\limits_{i=1}^{K}[\frac{1}{U_i}\sum\limits_{\vec{X_j}\in U_i}\max\limits_{\vec{X_q}\in U_i}\{d(\vec{X_j},\vec{X_q})\}]}{\sum\limits_{i=1}^{K}[\min\limits_{j\in k,j\neq i}\{d(\vec{m_i},\vec{m_j})\}]}
$$

$$(7.7)$$

where $d(\vec{X_j}, \vec{X_q})$ denotes the distance metric between two data points $\vec{X_j}$ and $\vec{X_q}$. It can be easily noted that this measure is a function of the ratio of the sum of within-cluster scatter to between-cluster separation. It is also observable that this has the same type of rational as the Dunn's index (DI) [314] and DB index [281]. The goodness of the valid optimal partition defines the smaller CS measure.

7.4 Automatic Clustering Differential Evolution (ACDE) Algorithm

Automatic clustering differential evolution (ACDE) algorithm [88] is an improved version of the classical differential evolution (DE) algorithm to cluster unlabeled datasets automatically. Firstly, the different parameters in DE have been tuned in ACDE for better functioning. The amplification factor of the difference vector (F) and the crossover control parameter (CCP) have been modified to improve the convergence properties. The amplification factor (F) is varied randomly in the range of (0.5, 1) by using the relation [88]

$$F = 0.5x(1 + rnd(0, 1))\qquad(7.8)$$

where, $rnd(0, 1)$ is a uniformly distributed random number within the range [0, 1]. The diversity of the population is maintained as the search progresses by tuning the crossover control parameter (CCP). It is decreased linearly with time from CCP_{max} = 1.0 to CCP_{min} = 0.5. All components of the parent vector are substituted by the difference vector operator. The value of the CCP decreases as the process progresses according to the following equation [88]

$$CCP = (CCP_{max} - CCP_{min})x(MNOI - cit)/MNOI\qquad(7.9)$$

where, CCP_{max} and CCP_{min} are the maximum and minimum values of the crossover control parameter (CCP), respectively and cit is the current iteration number and $MNOI$ is the maximum number of iterations.

The components of the parent vectors are substituted by the difference vector operator at the initial stages of the operation but the rate of substitution of the components becomes lesser as the process progresses.

The most important feature of this algorithm is the chromosome representation for the automatic determination of the optimal number of clusters in an unlabeled dataset. Here, the chromosome is devised as a vector of real numbers and the dimension of the chromosome is denoted as $K_{max} + (K_{max} \times D)$, where, K_{max} is the maximum number of clusters. The first K_{max} values are positive floating point numbers within [0, 1] and these are represented as the activation threshold value for each cluster centroid. The active/inactive cluster centroids are decided based on some predefined threshold values. A cluster is indicated as an active cluster if the threshold value of the corresponding cluster centroid is greater than 0.5, otherwise, it is designated as an inactive cluster. Thus, the optimal number of the active clusters is settled on the basis of the threshold value.

ACDE is efficient in determining the optimal number of clusters from datasets. However, the main limitation of this algorithm lies in the selection of an appropriate threshold for the cluster centroids. The choice of 0.5 as the threshold value is simply heuristic and hence far from being considered as a full-proof choice in diverse data distributions.

7.5 GA-Based Clustering Algorithm Validated by Fuzzy Intercluster Hostility Index

It is already mentioned that the classical DE algorithm is unable to determine the optimal number of clusters from an unlabeled dataset automatically. The number of target clusters is always provided as an input to the algorithm for arriving at an optimal clustered output. It is also clear that though the ACDE algorithm improves upon the classical DE algorithm in the automatic determination of optimal number of clusters from a starting value of the number of clusters, yet the choice of thresholding out of the inactive clusters is heuristic in nature.

In this section, we present an unsupervised method for the determination of the validity of the cluster centroids derived from a genetic algorithm based approach. The basic principle of this cluster validation process is centered on the underlying intercluster heterogeneity of the resultant clusters.

As is evident from the properties of the fuzzy intercluster hostility index discussed in Sect. 7.2.1, two clusters are denoted as homogeneous to each other if the fuzzy intercluster hostility index is close to zero (0). A value of the fuzzy intercluster hostility close to one (1) implies that two clusters are heterogeneous/hostile to each other. Given this property, a particular cluster can be merged with its neighbor if the clusters are homogeneous to each other. Thus, the objective of the cluster validation procedure using the fuzzy intercluster hostility index is to find out the distinct clusters, by eliminating some of the clusters through merging with their neighboring clusters by investigating the corresponding fuzzy intercluster hostility indices. The resultant clusters which remain distinct after the validation process are those ones maintaining appreciable heterogeneity between themselves.

The clustering process generally assigns a point (p_{ij}) to a particular cluster (out of K clusters) with cluster centroid m_{ij} firstly by calculating the distance $dt(p_{ij}, m_{ij})$ between a point (p_{ij}) in the image data/dataset and the cluster centroids (m_{ij}), and secondly by the following condition [319–323].

$$dt(p_{ij}, m_i j) = \text{Min}_{\forall c \in \{1,2,\dots,K\}}\{dt(p_{ij}, m_{ic})\} \qquad (7.10)$$

For an image dataset to be clustered into $K_{\max}$ number of regions $(R_1, \dots, R_{K_{\max}})$, the fuzzy intercluster hostility index $(\rho_{i,i+1}, 1 \leq i \leq K_{\max-1})$ between each and every other region holds the key for the target objective. The average fuzzy intercluster hostility index (ρ_{avg}) over all the possible regions reflects the average global heterogeneity. The validity of a cluster is decided by comparing its fuzzy intercluster hostility index with ρ_{avg}. The algorithm is described as follows [319–323].

```
1 Begin
2 Initialize Pop[iter], iter:=0
3 Cluster(Pop[iter])
```

Remark: **Pop[iter]** is the initial population of class boundaries required to cluster an image dataset.

```
4 Do
5 i:=0, Determine RHO[i][i+1], 1<=i<=Kmax-1
6 RHO_avg:= SUM(RHO[i][i+1])/Kmax, 1<=i<=Kmax-1
```

Remark: **RHO[i][i+1]** is the fuzzy intercluster hostility index between ith and $(i+1)th$ clusters. **RHO_avg** is the average fuzzy intercluster hostility index.

```
7 Sort all RHO[i][i+1]<=RHO_avg in ascending order
8 Do
9 Determine the smallest among RHO[i][i+1]<=RHO_avg
10 If RHO[i-1][i]<RHO[i+1][i+2] Then Omit R[i]
11 If RHO[i-1][i]>RHO[i+1][i+2] Then Omit R[i+1]
13 If RHO[1][2]<RHO_avg Then Omit R[1]
14 If RHO[Rmax-1][Rmax]<=RHO_avg Then Omit R[max]
```

Remark: Cluster centroids R_1 or $R_{\max}$ refer to the first and last possible centroids, respectively.

```
15 Loop Until RHO[i][i+1]>RHO[avg]
16 Compute Fit(Pop[iter])
17 iter:=iter+1
18 Select Pop[iter]
19 Crossover Pop[iter]
```

```
20 Mutate Pop[iter]
21 Loop Until iter=MaxGen
22 End
```

Hence, the valid number of clusters is determined by this algorithm by the omission of the redundant cluster centroids. The redundancy is decided on the basis of the fuzzy intercluster hostility index between the corresponding adjacent regions.

7.6 Results

In this section the presented method is compared with the ACDE [88] algorithm. All the methods have been applied on three gray level Peppers, Lena and Baboon images, each of dimension 128×128. In both of these clustering techniques, the basic feature of the image data is the intensity level of each pixels. The number of data items present in each image is 16384 though the datapoints of those images are single dimensional. The population size for both the processes is 100. The crossover probability (μ_c) and the mutation probability (μ_p)for the presented method are 0.9 and 0.01, respectively. The maximum (K_{max}) and the minimum (K_{min}) number of clusters as applied in the presented method and in the ACDE are 20 and 2, respectively. At the initial stage, the chromosome size is equal for both the processes. In ACDE algorithm, the control genes or the activation thresholds are generated randomly in the range [0, 1]. The cluster centroids are also randomly determined between maximum and minimum gray scale value of the test images. The parameters for these clustering algorithms are tabulated in the Table 7.1. The fuzzy intercluster hostility index based automatic clustering algorithm is denoted as FIH algorithm in the heading/caption of different tables and figures.

In this chapter, to determine the performance of the FIH based automatic clustering algorithm in comparison with the ACDE algorithm, it has been focused on two major issues: firstly, determination of the optimal number of clusters and for that, quality

Table 7.1 Parameters for the clustering algorithms

FIH algorithm		ACDE	
Parameter	Value	Parameter	Value
Pop_size	100	Pop_size	100
Crossover probability (μ_c)	0.9	CCR_{max}	1.0
Mutation probability (μ_p)	0.01	CCR_{min}	0.5
K_{max}	20	K_{max}	20
K_{min}	2	K_{min}	2

of the solution is evaluated by the DB and CS measures. For that purpose, we have conducted 20 independent runs of each algorithm. The ultimate results have been expressed in terms of the mean values and standard deviations over the 20 runs in each case. Among the 20 independent runs, six DB measure-based better results and the derived optimal number of clusters of the Peppers image are reported in Table 7.2. Similarly, six DB measure-based better results and the derived optimal number of clusters of the Lena and Baboon images are tabulated in Tables 7.3 and 7.4, respectively.

Table 7.2 DB measure and number of active clusters of Peppers Image

Sl. No.	FIH algorithm		ACDE	
	DB measure	# Active clusters	DB measure	# Active clusters
1	0.4785	5	0.6176	6
2	0.5050	5	0.7480	4
3	0.5061	6	0.5680	4
4	**0.4701**	**6**	0.5699	5
5	0.5302	7	**0.5582**	**5**
6	0.5301	6	0.6181	5

Table 7.3 DB measure and number of active clusters of Lena Image

Sl. No.	FIH algorithm		ACDE	
	DB measure	# Active clusters	DB measure	# Active clusters
1	0.5181	5	0.5222	5
2	**0.5122**	**5**	0.5950	5
3	0.5198	5	0.5264	4
4	0.5260	6	**0.5187**	**4**
5	0.5894	6	0.5575	5
6	0.5698	7	0.4839	4

Table 7.4 DB measure and number of active clusters of Baboon Image

Sl. No.	FIH Algorithm		ACDE	
	DB measure	# Active clusters	DB measure	# Active clusters
1	0.5309	6	0.5709	5
2	0.6041	6	0.6930	5
3	0.5989	6	0.5654	5
4	0.5594	5	0.6809	7
5	**0.5252**	**7**	0.5812	5
6	0.5754	6	**0.5224**	**6**

After that, the final fitness value, the number of clusters found, the intracluster distance and the intercluster distance are noted for these three test images in Table 7.5. The mean distance between data vectors within a cluster is known as intracluster distance and it is desirable to minimize the intracluster distances. On the contrary, the intercluster distance is measured as the mean distance between the centroids of the clusters and the maximization of the distance between the clusters is the objective. The mean and standard deviation of the number of classes found, the final DB measure value, the intracluster distance and the intercluster distance obtained for each algorithm are reported in the columns 3, 4, 5 and 6 of Table 7.5.

After that, the segmented test images by the fuzzy intercluster index based automatic image segmentation algorithm and ACDE algorithm are shown in Figs. 7.1, 7.2, 7.3, 7.4, 7.5 and 7.6. The DB measure acted as the fitness function to evaluate the quality of these segmented images. The segmented test images by the presented method are presented in Figs. 7.1, 7.2 and 7.3. The images in Figs. 7.4, 7.5 and 7.6 are ACDE algorithm based segmented test images. The number of segmentation classes (K) is mentioned in those images.

The same experiment has been repeated again with the CS measure. The CS measure-based fitness function, instead of DB index-based fitness function, along with all other components are reported in Tables 7.6, 7.7, 7.8 and 7.9. These tables are comparable with Tables 7.2, 7.3, 7.4 and 7.5 with respect to experimental results. The best entries are marked **boldfaced** in all the tables.

Like the DB measure based segmented outputs, the CS measure based segmented test images are presented in the Figs. 7.7, 7.8, 7.9, 7.10, 7.11 and 7.12. The segmented images exhibited in the Figs. 7.7, 7.8 and 7.9 are derived by the FIH based algorithm. The ACDE algorithm based segmented test images are demonstrated in Figs. 7.10, 7.11 and 7.12. The number of segmentation classes (K) is also mentioned in those images.

Table 7.5 Final solution (Mean and standard deviation over 20 independent runs) after each algorithm was terminated after running for 10^3 iterations with the DB measure-based fitness function

Image	Algorithm	Avg. no of Clusters found	DB measure	Mean intra cluster distance	Mean inter cluster distance
Peppers	FIH	**5.56 ± 0.7265**	**0.5129 ± 0.0357**	**12.2707 ± 3.654**	**210.142 ± 38.655**
	ACDE	5.00 ± 0.7071	0.6078 ± 0.0651	15.6336 ± 4.639	192.959 ± 41.489
Lena	FIH	**5.63 ± 0.7440**	**0.5404 ± 0.0273**	**9.8506 ± 1.642**	**180.207 ± 15.611**
	ACDE	4.75 ± 0.7071	0.5500 ± 0.0417	15.4641 ± 5.086	174.359 ± 34.877
Baboon	FIH	**5.67 ± 0.7071**	**0.5544 ± 0.0335**	**9.3074 ± 1.019**	**177.699 ± 32.723**
	ACDE	5.25 ± 0.8864	0.5906 ± 0.0663	11.8009 ± 2.823	174.140 ± 22.813

Fig. 7.1 DB measure based segmented Peppers image by FIH algorithm with **a** $K = 6$. **b** $K = 6$. **c** $K = 7$. **d** $K = 6$ classes

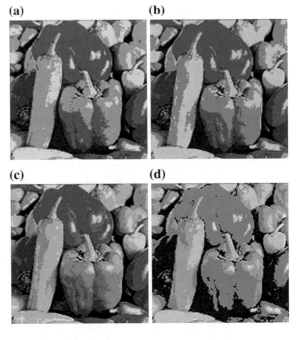

Fig. 7.2 DB measure based segmented Lena image by FIH algorithm with **a** $K = 6$. **b** $K = 7$. **c** $K = 6$. **d** $K = 6$ classes

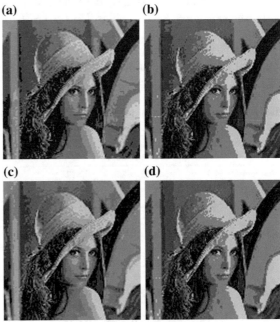

Fig. 7.3 DB measure based segmented Baboon image by FIH algorithm with **a** $K = 7$. **b** $K = 6$. **c** $K = 6$. **d** $K = 6$ classes

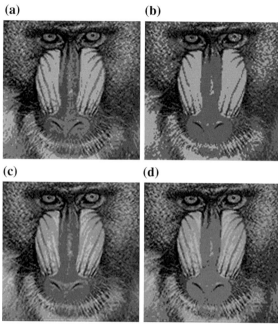

Fig. 7.4 DB measure based segmented Peppers image by ACDE algorithm with **a** $K = 6$. **b** $K = 5$. **c** $K = 5$. **d** $K = 6$ classes

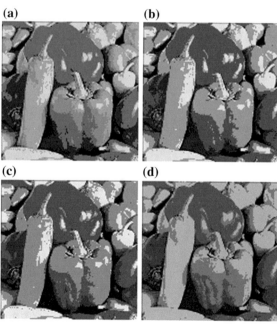

Fig. 7.5 DB measure based segmented Lena image by ACDE algorithm with **a** $K = 5$. **b** $K = 5$. **c** $K = 6$. **d** $K = 5$ classes

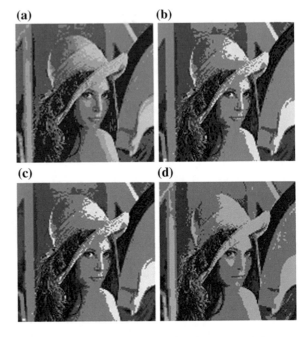

Fig. 7.6 DB measure based segmented Baboon image by ACDE algorithm with a $K = 5$. **b** $K = 5$. **c** $K = 7$. **d** $K = 6$ classes

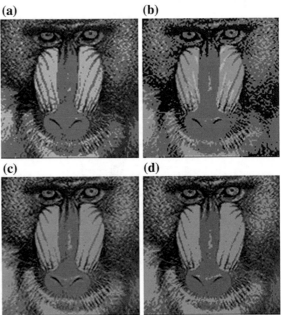

Table 7.6 CS measure and number of active clusters of Peppers Image

Sl. No.	FIH Algorithm		ACDE	
	CS measure	# Active clusters	CS measure	# Active clusters
1	0.9383	6	0.9973	6
2	0.9665	7	0.9791	7
3	0.9616	7	0.8891	9
4	0.9488	7	0.9934	8
5	0.9271	7	**0.8304**	6
6	**0.9219**	6	1.0369	7

Table 7.7 CS measure and number of active clusters of Lena Image

Sl. No.	FIH Algorithm		ACDE	
	CS measure	# Active clusters	CS measure	# Active clusters
1	0.9890	6	1.0353	7
2	0.9356	8	0.9520	7
3	0.9819	7	0.9881	6
4	0.9752	6	**0.9416**	8
5	0.9319	9	1.0476	5
6	**0.8971**	7	0.9832	6

Table 7.8 CS measure and number of active clusters of Baboon Image

Sl. No.	FIH Algorithm		ACDE	
	CS measure	# Active clusters	CS measure	# Active clusters
1	0.9579	5	0.9877	8
2	0.9667	6	0.9578	6
3	0.9027	7	**0.9002**	5
4	0.9073	7	0.9110	6
5	**0.8672**	8	0.9453	8
6	0.9486	5	0.9881	7

Not only confined within the automatic clustering of the single-dimensional real-life images, the automatic clustering of the real-life multidimensional data set is also tried by fuzzy intercluster hostility index based clustering algorithm. Three real-life multidimensional data sets [6, 328], i.e., iris, vowel and glass data are used in this chapter. In this regard, few conventions are used, such as, n is the number of data points, K is the number of clusters and d is the number of features.

- **Iris data set**: This is widely known database with $n = 150$ data vectors, $d = 4$ inputs, and $K = 3$ classes. Basically, it is a database of iris flower with three different species: *Iris setosa*, *Iris virginica* and *Iris versicolor*. Each cluster has 50

Table 7.9 Final solution (Mean and standard deviation over 20 independent runs) after each algorithm was terminated after running for 10^3 iterations with the CS measure-based fitness function

Image	Algorithm	Avg. no of Clusters found	CS measure	Mean Intra Cluster Distance	Mean Inter Cluster Distance
Peppers	FIH	**7.05 ± 1.8340**	**0.9403 ± 0.2838**	**11.7372 ± 2.013**	**265.672 ± 86.767**
	ACDE	7.08 ± 2.6628	0.9514 ± 0.2995	14.4351 ± 6.066	263.358 ± 93.256
Lena	FIH	**7.00 ± 1.1180**	**0.9629 ± 0.0333**	**14.3944 ± 5.501**	**271.636 ± 28.343**
	ACDE	7.00 ± 1.3093	0.9683 ± 0.0604	15.8177 ± 2.164	264.635 ± 50.184
Baboon	FIH	**6.00 ± 1.0954**	**0.9446 ± 0.0356**	**17.5644 ± 5.716**	**205.312 ± 41.912**
	ACDE	7.00 ± 1.0836	0.9817 ± 0.0701	18.5943 ± 3.729	201.767 ± 26.143

Fig. 7.7 CS measure based segmented Peppers image by FIH algorithm with **a** $K = 6$. **b** $K = 6$. **c** $K = 7$. **d** $K = 6$ classes

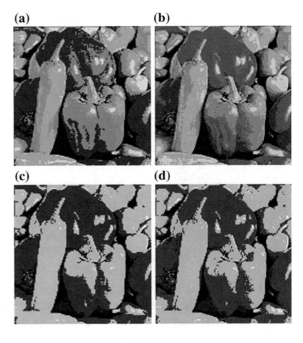

Fig. 7.8 CS measure based segmented Lena image by FIH algorithm with **a** $K = 6$. **b** $K = 6$. **c** $K = 8$. **d** $K = 9$ classes

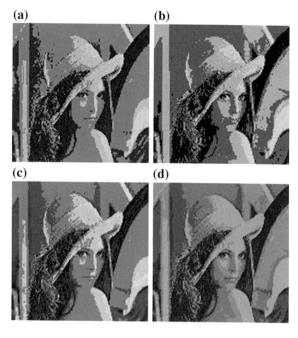

Fig. 7.9 CS measure based segmented Baboon image by FIH algorithm with **a** $K = 7$. **b** $K = 7$. **c** $K = 8$. **d** $K = 5$ classes

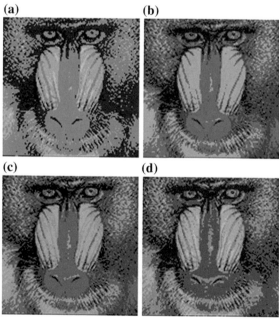

Fig. 7.10 CS measure based segmented Peppers image by ACDE algorithm with **a** $K=$ 8. **b** $K=9$. **c** $K=6$. **d** $K=7$ classes

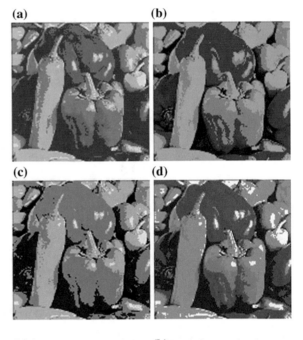

Fig. 7.11 CS measure based segmented Lena image by ACDE algorithm with **a** $K=$ 7. **b** $K=7$. **c** $K=8$. **d** $K=6$ classes

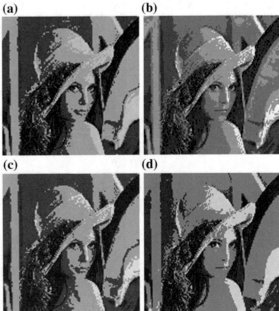

Fig. 7.12 CS measure based segmented Baboon image by ACDE algorithm with **a** $K = 8$. **b** $K = 6$. **c** $K = 7$. **d** $K = 5$ classes

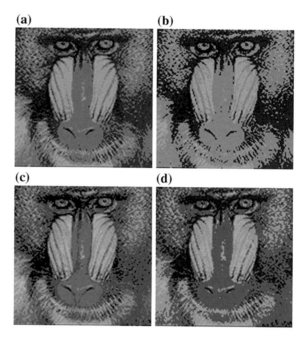

number of objects as each species consists of 50 samples. Each sample has four features, such as, sepal length, sepal width, petal length, and petal width.

- **Vowel data set**: This data set comprises $n = 871$ Indian Telugu vowel sounds with $d = 3$ inputs, and $K = 6$ classes. Individual data vector has three features, viz. F_1, F_2, and F_3, viz. to the first, second and, third vowel frequencies, and six overlapping classes d (72 objects), a (89 objects), i (172 objects), u (151 objects), e (207 objects), o (180 objects).

- **Glass data set**: This data set has $n = 214$ data vectors, $d = 9$ inputs, and $K = 6$ classes. Glass data set is classified into six different types of glasses: (i) building windows float processed (70 objects); (ii) building windows nonfloat processed (76 objects); (iii) vehicle windows float processed (17 objects); (iv) containers (13 objects); (v) tableware (9 objects); and (vi) headlamps (29 objects). Each data vector has nine features: (i) refractive index; (ii) sodium; (iii) magnesium; (iv) aluminum; (v) silicon; (vi) potassium; (vii) calcium; (viii) barium; and (ix) iron.

Like the image data, automatic clustering of these multidimensional data set is done by the FIH based algorithm and the ACDE algorithm and the performance of these algorithms are evaluated in terms of DB and CS measures. Determination of the optimal number of clusters and the evaluation of the quality of solution are the two issues that will be the outcome from this experiment. To do that, same parameter values, those are reported in Table 7.1, are applied in these two algorithms. Each algorithm are run independently for 40 times to find out the final solutions. Six DB measure-based better results and the corresponding derived optimal number of clusters of these data sets are reported in the Tables 7.10, 7.11 and 7.12. The ultimate results have been expressed in terms of the mean values and standard deviations over the 40 runs in each case and they are shown in Table 7.13. The mean and standard deviation of the intracluster distance between data vectors and the mean and standard deviation of the intercluster distance between the cluster centroids of these data sets are also reported in Table 7.13.

All the experiments are acquitted again for the former group of Tables 7.10, 7.11, 7.12 and 7.13. This time, the experimental results are derived on the basis of the CS measure which acts as the fitness function. In Tables 7.14, 7.15 and 7.16, the CS measure-based fitness function and the derived number of active clusters are reported and the best results of each algorithm are marked in **boldface**. The mean and standard deviation of the CS measure-based fitness function, the number of active clusters, intracluster distance and intercluster distance are tabulated in Table 7.17. The better results are also labeled **boldfaced** in that table.

A detailed scrutiny of different tables reveals that the FIH based algorithm for the automatic clustering of data set outperforms the ACDE algorithm. The comparison will be discussed one by one. From the Tables 7.2 and 7.5, it is observed that the DB measure value for the Peppers image derived by the FIH based algorithm is better than the same value evolved by the ACDE algorithm. Same thing also happens for the Lena image if we go through Tables 7.3 and 7.5. In Table 7.4, we find that it is only one case (for the Baboon image) that the best result is evaluated by the ACDE

Table 7.10 DB measure and number of active clusters of Iris data

Sl. No.	FIH algorithm		ACDE	
	DB measure	# Active clusters	DB measure	# Active clusters
1	0.6209	3	0.9156	3
2	0.6846	3	0.9560	3
3	0.6154	3	0.9047	3
4	0.6401	3	0.9039	3
5	0.6583	3	0.9095	3
6	**0.5911**	**3**	**0.8532**	**3**

Table 7.11 DB measure and number of active clusters of Vowel data

Sl. No.	FIH algorithm		ACDE	
	DB measure	# Active clusters	DB measure	# Active clusters
1	0.8053	6	1.0351	6
2	0.7878	6	**1.0118**	**5**
3	0.8351	6	1.0658	6
4	0.8240	6	1.2941	6
5	**0.7701**	**6**	1.262	6
6	0.8947	6	1.0219	6

Table 7.12 DB measure and number of active clusters of Glass data

| Sl. No. | FIH algorithm | | ACDE | |
	DB measure	# Active clusters	DB measure	# Active clusters
1	0.7929	6	1.5719	6
2	**0.7793**	**6**	1.5837	6
3	0.8107	6	1.6021	5
4	0.7956	6	1.6013	5
5	0.8003	6	1.6010	5
6	0.8267	6	**1.5594**	**5**

Table 7.13 Final solution (Mean and standard deviation over 40 independent runs) after each algorithm was terminated after running for 10^5 iterations with the DB measure-based fitness function

Data set	Algorithm	Avg. no of clusters found	DB measure	Mean intra cluster distance	Mean inter cluster distance
Iris data	FIH	**3.05 ± 0.2236**	**0.6759 ± 0.0443**	**0.9038 ± 0.1766**	**4.375 ± 0.9135**
	ACDE	3.45 ± 0.6048	0.9354 ± 0.0441	1.392 ± 0.2413	3.968 ± 0.8082
Vowel data	FIH	**6.06 ± 0.5737**	**0.8446 ± 0.0395**	**238.458 ± 23.387**	**979.819 ± 49.625**
	ACDE	5.60 ± 0.7367	1.0973 ± 0.1128	330.242 ± 26.989	947.785 ± 65.316
Glass data	FIH	**6.17 ± 0.3929**	**0.8169 ± 0.0384**	**1.325 ± 0.0609**	**15.861 ± 1.543**
	ACDE	5.15 ± 0.3755	1.58 ± 0.0352	3.965 ± 0.1394	13.849 ± 1.627

Table 7.14 CS measure and number of active clusters of Iris data

| Sl. No. | FIH Algorithm | | ACDE | |
	CS measure	# Active clusters	CS measure	# Active clusters
1	**0.6616**	**3**	0.7316	3
2	0.6805	3	0.6647	3
3	0.6938	3	0.7242	3
4	0.6730	3	0.6707	3
5	0.6688	3	**0.6622**	**3**
6	0.6805	3	0.6882	4

Table 7.15 CS measure and number of active clusters of Vowel data

Sl. No.	FIH algorithm		ACDE	
	CS measure	# Active clusters	CS measure	# Active clusters
1	0.8791	6	1.2821	7
2	0.8979	6	1.2984	5
3	0.8911	6	1.2935	6
4	0.8957	6	**1.1224**	**5**
5	**0.8729**	**6**	1.2659	5
6	0.8856	6	1.2993	5

Table 7.16 CS measure and number of active clusters of Glass data

Sl. No.	FIH Algorithm		ACDE	
	CS measure	# Active clusters	CS measure	# Active clusters
1	**0.7806**	**6**	**1.0437**	**6**
2	0.8591	6	1.2070	5
3	0.8148	6	1.2594	6
4	0.8646	6	1.2175	6
5	0.8687	6	1.2971	6
6	0.8581	6	1.2165	6

Table 7.17 Final solution (Mean and standard deviation over 40 independent runs) after each algorithm was terminated after running for 10^5 iterations with the CS measure-based fitness function

Data set	Algorithm	Avg. no of clusters found	CS measure	Mean intra cluster distance	Mean inter cluster distance
Iris data	FIH	**3.06 ± 0.2581**	**0.6808 ± 0.0130**	**1.053 ± 0.1660**	**4.511 ± 0.852**
	ACDE	3.66 ± 0.4879	0.6912 ± 0.0237	1.523 ± 0.3569	3.651 ± 0.812
Vowel data	FIH	**6.13 ± 0.6399**	**0.8879 ± 0.0098**	**289.269 ± 50.630**	**974.141 ± 62.751**
	ACDE	5.87 ± 1.7265	1.2066 ± 0.0868	355.984 ± 36.461	952.527 ± 58.599
Glass data	FIH	**6.00 ± 0.4851**	**0.8379 ± 0.0398**	**1.572 ± 0.5254**	**16.035 ± 1.607**
	ACDE	5.29 ± 0.6112	1.1949 ± 0.0562	3.899 ± 0.664	12.871 ± 2.635

algorithm. The mean and standard deviation of DB measure for the Baboon image in Table 7.5 reveals that the FIH based algorithm outperforms the ACDE algorithm. In the same way, we have conducted our study with the CS measure also and the results are tabulated in Tables 7.6, 7.7, 7.8 and 7.9. The results in those tables reveal that the FIH based algorithm generates better CS measure value for all the test images than the ACDE algorithm. Interestingly, we found from Tables 7.5 and 7.9 that the mean intracluster distance and mean intercluster distance for all the test images obtained by the FIH based algorithm are much better than the ACDE algorithm. This is also evident from the segmented outputs shown in Figs. 7.1, 7.2, 7.3, 7.4, 7.5, 7.6, 7.7, 7.8, 7.9, 7.10, 7.11 and 7.12. The segmented test images are more prominent than the segmented test images reported by the ACDE algorithm. The FIH based algorithm outperformed the ACDE algorithm in respect of the segmentation quality of the images for the different number of classes.

Now we will concentrate on the real life multidimensional data sets. For the iris data set, the DB measure deduced by the FIH based algorithm is significantly better than the same value gained by the ACDE algorithm and this can be concluded from the Tables 7.10 and 7.13. From the Tables 7.11 and 7.13, it is learnt that the DB measure value for the vowel data set obtained by the FIH based algorithm outperforms the ACDE algorithm. We can find out from the Tables 7.12 and 7.13 that the FIH based algorithm derives better DB measure value than the ACDE algorithm for the glass data set. In fact, the CS measure for different data sets derived by the FIH based algorithm offers better result than ACDE algorithm and this thing is revealed from the Tables 7.14, 7.15, 7.16 and 7.17. It is also evident from Tables 7.13 and 7.17 that the mean intracluster distance and mean intercluster distance for all the multidimensional data sets found by the FIH based algorithm are much better than the ACDE algorithm.

At the end we can conclude that the FIH based algorithm performs better segmentation/clustering than the ACDE algorithm.

7.7 Discussions and Conclusion

A cluster validation approach for the determination of the optimal number of clusters in a dataset is presented in this chapter. The basis of this validation procedure for distinguishing valid and invalid optimal clusters, derived from a genetic algorithm, lies in measuring the fuzzy intercluster hostility index to reflect the underlying heterogeneity between the clusters. The FIH based approach has been successful in determining the valid optimal clusters from gray level images as well as multidimensional datasets. Comparative results are reported with reference to the clusters obtained with the automatic clustering differential evolution (ACDE) algorithm.

Both the methods are applied on different types of test images as well as different type of multidimensional datasets characterized by multiple number of features. It is proved that the FIH based algorithm outperforms the ACDE algorithm in all respects.

References

1. L.A. Zadeh, Fuzzy sets. Inf. Control **8**(3), 338–353 (1965)
2. L.A. Zadeh, Outline of a new approach to analysis of complex systems and decision processes. IEEE Trans. Syst. Man Cybern. **3**, 28–44 (1973)
3. A. Ghosh, N.R. Pal, S.K. Pal, Self-organization for object extraction using a multilayer neural network and fuzziness measures. IEEE Trans. Fuzzy Syst. **1**(1), 54–68 (1993)
4. R. Krishnapuram, J. Keller, The possibilistic c-means algorithm: Insights and recommendations. IEEE Trans. Fuzzy Syst. **4**(3), 385–393 (1996)
5. K. Krishna, M.N. Murty, Genetic K-means algorithm. IEEE Trans. Syst. Man Cybern. Part B: Cybern. **29**, 433–439 (1999)
6. S.K. Pal, D.D. Majumder, Fuzzy sets and decision making approaches in vowel and speaker recognition, *IEEE Transactions on Systems, Man, Cybernatics*, vol. SMC-7, pp. 625–629 (1977)
7. C. Carson, S. Belongie, H. Greenspan, J. Malik, Blobworld: image segmentation using expectation-maximization and its application to image querying. IEEE Trans. Pattern Anal. Mach. Intell. **24**(8), 1026–1038 (2002)
8. L.A. Vese, T.F. Chan, A multiphase level set framework for image segmentation using the mumford and shah model. Int. J. Comput. Vis. **50**(3), 271–293 (2002)
9. P.F. Felzenszwalb, D.P. Huttenlocher, Efficient graph-based image segmentation. Int. J. Comput. Vis. **59**(2), 167–181 (2004)
10. R.I. Kitney, N. Smith, *X-ray Reflective Lens Arrangement* (2005)
11. Y.B. Chen, O.T.-C. Chen, *Image segmentation method using thresholds automatically determined from picture contents* (EURASIP J, Image Video Process, 2009)
12. S.M. Praveena, I. Vennila, Optimization fusion approach for image segmentation using K-means algorithm. Int. J. Comput. Appl. **2**(7) (2010)
13. N. Sharma, L.M. Aggarwal, Automated medical image segmentation techniques. J. Med. Phys. **35**(1), 3–14 (2010)
14. P. Arbelaez, M. Maire, C. Fowlkes, J. Malik, Contour detection and hierarchical image segmentation. IEEE Trans. Pattern Anal. Mach. Intell. **33**(5), 898–916 (2011)
15. S. Jamil, F. Guimares, J. Cousty, Y. Kenmochi, L. Najman, *A Hierarchical Image Segmentation Algorithm Based on an Observation Scale* (Springer, Berlin, Heidelberg, 2012)
16. B. Peng, L. Zhang, D. Zhang, A survey of graph theoretical approaches to image segmentation. Pattern Recogn. **46**(3), 1020–1038 (2013)
17. N.R. Pal, S.K. Pal, A review on image segmentation techniques. Pattern Recogn. **26**(9), 1277–1294 (1993)
18. G. Iannizzotto, L. Vitta, Fast and accurate edge-based segmentation with no contour smoothing in 2-d real images. IEEE Trans. Image Process. **9**(7), 1232–1237 (2000)

© Springer International Publishing AG 2016
S. De et al., *Hybrid Soft Computing for Multilevel Image and Data Segmentation*, Computational Intelligence Methods and Applications, DOI 10.1007/978-3-319-47524-0

19. M. Brejl, M. Sonka, Object localization and border detection criteria design in edge-based image segmentation: automated learning from examples. IEEE Trans. Med. Imaging **19**(10), 973–985 (2000)
20. D.L. Pham, C. Xu, J.L. Prince, Current methods in medical image segmentation. Ann. Rev. Biomed. Eng. **2**, 315–337 (2000)
21. C. Kim, J.-N. Hwang, Fast and automatic video object segmentation and tracking for content-based applications. IEEE Trans. Circ. Syst. Video Technol. **12**(2), 122–129 (2002)
22. A. Tsai, A. Yezzi, W. Wells, C. Tempany, D. Tucker, A. Fan, W.E. Grimson, A. Willsky, A shape-based approach to the segmentation of medical imagery using level sets. IEEE Trans. Med. Imaging **22**(2), 137–154 (2003)
23. N. Senthilkumaran, R. Rajesh, Edge detection techniques for image segmentation-a survey of soft computing approaches. Int. J. Recent Trends Eng. Technol. **01**(2), 250–254 (2009)
24. T. Uemura, G. Koutaki, K. Uchimera, Image segmentation based on edge detection using boundary code. Int. J. Innovative Comput. Inf. Control **07**(10), 6073–6083 (2011)
25. N. Yokoya, M.D. Levine, Range image segmentation based on differential geometry: a hybrid approach. IEEE Trans. Pattern Anal. Mach. Intell. **11**(6), 643–649 (1989)
26. J.C. Bezdek, L.O. Hall, L.P. Clarke, Review of MR image segmentation techniques using pattern recognition. Med. Phys. **20**(4), 1033–1048 (1993)
27. K. Haris, S.N. Efstratiadis, N. Maglaveras, A.K. Katsaggelos, Hybrid image segmentation using watersheds and fast region merging. IEEE Trans. Image Process. **7**(12), 1684–1699 (1998)
28. M. Piccardi, Background subtraction techniques: a review. IEEE Int. Conf. Syst. Man Cybern. no. 0-7803-8566-7/04, 3099–3104 (2004)
29. K.S. Tan, N.A.M. Isa, Color image segmentation using histogram thresholding-fuzzy C-means hybrid approach. Pattern Recogn. **44**(1), 1–15 (2011)
30. M. Erdt, S. Steger, G. Sakas, Regmentation: a new view of image segmentation and registration. J. Radiat. Oncol. Inf. **4**(1), 1–23 (2012)
31. T.V. Spina, P.A.V. de Miranda, A.X. Falco, Hybrid approaches for interactive image segmentation using the live markers paradigm. IEEE Trans. Image Process. **23**(12), 5756–5769 (2014)
32. S. Bhattacharyya, P. Dutta, S. Chakraborty (eds.), vol. 611 (Springer India, 2016)
33. S. De, S. Bhattacharyya, *Color Magnetic Resonance Brain Image Segmentation by ParaOptiMUSIG Activation Function: An Application*, vol. 611 (Springer India, 2016)
34. S. Ganguly, D. Bhattacharjee, M. Nasipuri, *Hybridization of 2D-3D Images for Human Face Recognition*, vol. 611 (Springer India, 2016)
35. U.R. Gogoi, M.K. Bhowmik, D. Bhattacharjee, A.K. Ghosh, G. Majumdar, *A Study and Analysis of Hybrid Intelligent Techniques for Breast Cancer Detection Using Breast Thermograms*, vol. 611 (Springer India, 2016)
36. http://commons.wikimedia.org/wiki/File:Brain_MRI.jpg
37. S. Haykin, *Neural networks: A comprehensive foundation* (Upper Saddle River, NJ, Prentice Hall, second ed. 1999)
38. J. Hertz, A. Krogh, R.G. Palmer, *Introduction to the Theory of Neural Computation* (Addison-Wesley, Reading, MA, 1994)
39. T. Kohonen, *Self-Organization and Associative Memory* (Springer, Berlin, 1989)
40. J.M. Zurada, *Introduction to Artificial Neural Syatems* (West Publishing Company, New York, 1992)
41. S. Mitra, T. Acharya, *Data Mining: Multimedia, Soft Computing and Bioinformatics* (Wiely-Intescience, Wiely, Hoboken, New Jersey, 2003)
42. T. Kohonen, Self-organized formation of topologically correct feature maps. Biol. Cybern. **43**, 59–69 (1982)
43. T. Kohonen, Self-organizing maps. Springer Series in Information Sciences, vol. 30 (1995)
44. L. Fausett, *Fundamentals of Neural Networks: Architectures* (Algorithms and Applications, Pearson Education, 1994)

45. S. Bhattacharyya, U. Maulik, *Soft Computing-Image and Multimedia Data Processing* (Springer, Heidelberg, Germany, 2013)
46. S. Kumar, *Neural Networks: A Classroom Approach* (Tata McGraw-Hill, New Delhi, 2004)
47. R. Rojas, *Neural Networks: A Systematic Introduction* (Springer, Berlin, 1996)
48. S. Bhattacharyya, Neural networks: evolution, topologies, learning algorithms and applications, in *Cross-Disciplinary Applications of Artificial Intelligence and Pattern Recognition: Advancing Technologies*, eds. by V. Mago, N. Bhatia (Hershey, PA, USA, IGI Global, 2012)
49. S. Bhattacharyya, *Object Extraction in a Soft Computing Framework*. Ph.D. thesis (Jadavpur University, India, 2007)
50. D. de Ridder, M. Egmont-Petersen, H. Handels, Image processing with neural networks-a review. Pattern Recogn. **35**(10), 2279–2301 (2002)
51. W.S. McCulloch, W. Pitts, A logical calculus of the ideas immanent in nervous activity. Bull. Math. Biophys. **5**, 115–133 (1943)
52. M.D. Buhmann, *Radial Basis Functions: Theory and Implementations* (Cambridge University Press, 2003)
53. V. Vapnik, *The Nature of Statistical Learning Theory* (Springer, London, 1995)
54. V. Vapnik, S. Golowich, A. Smola, Support vector method for function approximation, regression estimation, and signal processing, in *Advances in Neural Information Processing Systems*, vol. 9 eds. by M. Mozer, M. Jordan, T. Petsche (Cambridge, MA, MIT Press, 1997), pp. 281–287
55. J.A. Anderson, *Introduction to Neural Networks* (MIT Press, Cambridge, MA, 1995)
56. K. Hornik, Approximation capabilities of multilayer feedforward networks. Neural Netw. **4**(2), 251–257 (1991)
57. J.J. Hopfield, Neural networks and physical systems with emergent collective computational abilities. Proc. National Acad. Sci. USA **79**, 2554–2558 (1982)
58. J.A. Anderson, J.W. Silverstein, S.A. Ritz, R.S. Jones, Distinctive features, categorical perception, probability learning: some applications of a neural model. Psychol. Rev. **84**, 413–451 (1977)
59. D.H. Ackley, G.E. Hinton, T.J. Sejnowski, A learning algorithm for boltzmann machines. Cogn. Sci. **9**, 147–169 (1985)
60. B. Kosko, Bidirectional associative memories. IEEE Trans. Systems Man Cybern. **18**(1), 49–60 (1988)
61. G.A. Carpenter, S. Grossberg, A massively parallel architecture for a self-organizing neural pattern recognition machine. Comput. Vis. Graph. Image Process. **37**, 54–115 (1987)
62. T.J. Ross, *Fuzzy Logic with Engineering Applications* (John Wiley, 2004)
63. L.A. Zadeh, Fuzzy logic and neural networks and soft computing. Commun. ACM **37**, 77–84 (1994)
64. T. Ross, J. Booker, J. Parkinson, *Fuzzy Logic and Probability Applications: Bridging the Gap* (Society for Industrial and Applied Mathematics, Philadelphia, PA, 2003)
65. B. Kosko, *Fuzzy Engineering* (Prentice Hall, Upper Saddle River, NJ, 1997)
66. S. Bhattacharyya, P. Dutta, Fuzzy logic: concepts, system design and applications to industrial informatics, in *Handbook of Research on Industrial Informatics and Manufacturing Intelligence: Innovations and Solutions*, eds. by M.A. Khan, A.Q. Ansari (Hershey, PA, USA, IGI Global, 2012), pp. 33–71
67. E. Cox, *The Fuzzy Systems Handbook* (Academic Press, Cambridge, MA, 1994)
68. D. Dubois, H. Prade, *Fuzzy Sets and Systems: Theory and Applications* (Academic Press, New York, 1980)
69. L.A. Zadeh, PRUF-a meaning representation language for natural languages. Int. J. Man-Mach. Stud. **10**, 395–460 (1978)
70. L.A. Zadeh, The concept of linguistic variable and its application to approximate reasoning: Part 1, 2 and 3. Inf. Sci. **8, 8, 9**, 199–249, 301–357, 43–80 (1975)
71. L.A. Zadeh, Fuzzy sets as a basis for a theory of possibility. Fuzzy Sets Syst. **1**, 3–28 (1978)
72. S. Bhattacharyya, P. Dutta, U. Maulik, Self organizing neural network (SONN) based gray scale object extractor with a multilevel sigmoidal (musig) activation function. Int. J. Found. Comput. Decis. Sci. **33**(2), 46–50 (2008)

73. D.E. Goldberg, *Genetic Algorithms: Search, Optimization and Machine Learning* (Addison-Wesley, New York, 1989)
74. L. Davis (ed.), *Handbook of Genetic Algorithms* (Van Nostrand Reinhold, New York, 1991)
75. H. Kargupta, K. Deb, D.E. Goldberg, Ordering genetic algorithms and deception, in *Proceedings of Parallel Problem Solving from Nature*, eds. by R. Manner, B. Manderick (Amsterdam, North-Holland, 1992), pp. 47–56
76. N.J. Radcliffe, Genetic set recombination, in *Foundations of Genetic Algorithms*, vol. 2, ed. by L.D. Whitley (Morgan Kaufmann, San Mateo, 1993), pp. 203–219
77. J.J. Grefenstette, R. Gopal, B. Rosmaita, D. Van Gucht, Genetic algorithms for the traveling salesman problem, in *Proceedings of 1st International Conference on Genetic Algorithms*, ed. by J.J. Grefenstette (Lawrence Erlbaum Associates, Hillsdale, 1985), pp. 160–168
78. J. Holland, *Adaptation in Neural Artificial Systems, Technical Report* (University of Michigan, Ann Arbor, MI, 1975)
79. D.E. Goldberg, K. Deb, B. Korb, Messy genetic algorithms: motivation, analysis, and first results. Complex Syst. **3**, 493–530 (1989)
80. D.E. Goldberg, K. Deb, B. Korb, Do not worry, be messy, in *Proceedings of 4th International Conference on Genetic Algorithms*, eds. by R.K. Belew, J.B. Booker (San Mateo, Morgan Kaufmann, 1991), pp. 24–30
81. D.E. Goldberg, K. Deb, H. Kargupta, G. Harik, Rapid, accurate optimization of difficult problems using fast messy genetic algorithms, in *Proceedings of 5th International Conference on Genetic Algorithms*, ed. by S. Forrest (Morgan Kaufmann, San Mateo, 1993), pp. 56–64
82. M. Paulinas, A. Uinskas, A survey of genetic algorithms applications for image enhancement and segmentation. Inf. Technol. Control **36**, 278–284 (2007)
83. R. Graham, *Pathfinding in Computer Games*. Proceedings of ITB Journal, 2004
84. C. Aranha, Portfolio management by genetic algorithms with error modeling. Inf. Sci. 459–465 (2007)
85. R. Storn, K. Price, Differential evolution-a simple and efficient heuristic for global optimization over continuous spaces. J. Global Optim. **11**, 341–359 (1997)
86. A.K. Qin, V.L. Huang, P.N. Suganthan, Differential evolution algorithm with strategy adaptation for global numerical optimization. IEEE Tran. Evol. Comput. **13**(2), 398–417 (2009)
87. A.K. Jain, M.N. Murty, P.J. Flynn, Data clustering: a review. ACM Comput. Surv. **31**(3), 264–323 (1999)
88. S. Das, A. Abraham, A. Konar, Automatic clustering using an improved differential evolution algorithm. IEEE Trans. Syst. Man Cybern. Part A: Syst. Humans **38**(1), 218–237 (2008)
89. N. Grira, M. Crucianu, N. Boujemaa, Unsupervised and semi-supervised clustering: a brief survey, in *A Review of Machine Learning Techniques for Processing Multimedia Content (Report of the MUSCLE European Network of Excellence (FP6)* (2004)
90. A. Demiriz, K. Bennett, M. Embrechts, Semi-supervised clustering using genetic algorithms. Artif. Neural Netw. Eng. (ANNIE-99) (1999), pp. 809–814
91. S. Basu, A. Banerjee, R.J. Mooney, Semisupervised clustering by seeding, in *Proceedings of 19th International Conference on Machine Learning (ICML-2002)* (2002), pp. 19–26
92. K. Wagstaff, C. Cardie, Clustering with Instance-level Constraints, in *Proceedings of the 17th International Conference on Machine Learning* (2000), pp. 1103–1110
93. R.C. Gonzalez, R.E. Woods, *Digital Image Processing* (Prentice-Hall, Englewood Cliffs, NJ, 2002)
94. M.R. Banham, A.K. Katsaggelos, Digital image restoration. IEEE Sig. Process. Mag. **14**(2), 24–41 (1997)
95. P.F. Felzenszwalb, D.P. Huttenlocher, Efficient graph-based image segmentation. Int. J. Comput. Vis. **59**(2), 167–181 (2004)
96. M. Sonka, V. Hlavac, R. Boyle, *Image Processing, Analysis and Machine Vision*, 2nd edn. (PWS Publishing, New York, 1999)
97. S. Bandyopadhyay, S. Saha, *Unsupervised Classification* (Springer, Berlin, Heidelbarg, 2013)
98. K. Deb, *Multi-Objective Optimization using Evolutionary Algorithms* (Wiley, Chichester, UK, 2001)

99. S. De, S. Bhattacharyya, S. Chakraborty, Multilevel Image Segmentation by a Multiobjective Genetic Algorithm Based OptiMUSIG Activation Function, in *Handbook of Research on Computational Intelligence for Engineering, Science and Business*, vol. 1, eds. by S. Bhattacharyya, P. Dutta, (Hershey, PA, USA, IGI Global, 2012), pp. 122–162
100. S. De, S.B.S. Chakraborty, B.N. Sarkar, P.K. Prabhakar, S. Bose, Gray Scale Image Segmentation by NSGA-II based OptiMUSIG Activation Function, in *International Conference on Communication Systems and Network Technologies (CSNT 2012)* (2012), pp. 104–108
101. K. Deb, S. Agrawal, A. Pratap, T. Meyarivan, A Fast Elitist Non-dominated Sorting Genetic Algorithm for Multi-objective Optimization: NSGA-II, in *Proceedings of the Parallel Problem Solving from Nature (PPSN VI)* (2000), pp. 849–858
102. K. Deb, A. Pratap, S. Agarwal, T. Meyarivan, A fast and elitist multiobjective genetic algorithm: NSGA-II. IEEE Trans. Evol. Comput. **6**(2), 182–197 (2002)
103. S. Bandyopadhyay, U. Maulik, A. Mukhopadhyay, Multiobjective genetic clustering for pixel classification in remote sensing imagery. IEEE Trans. Geosci. Remote Sens. **45**, 1506–1511 (2007)
104. C.C. Coello, Evolutionary multiobjective optimization: a historical view of the field. IEEE Comput. Intell. Mag. **1**(1), 28–36 (2002)
105. I.F. Sbalzariniy, S. Müllery, P. Koumoutsakos, Multiobjective optimization using evolutionary algorithms, in *Proceedings of the Summer Program on Center for Turbulence Research* (Thessaloniki, Greece, 2000), pp. 63–74
106. A. Mukhopadhyay, U. Maulik, A multiobjective approach to MR brain image segmentation. Appl. Soft Comput. **11**(1), 872–880 (2011)
107. K. Deb, Multi-objective genetic algorithms: problem difficulties and construction of test problems. Evol. Comput. **7**(3), 205–230 (1999)
108. J.D. Schaffer, *Multiple Objective Optimizations with Vector Evaluated Genetic Algorithms*. Ph.D. thesis (Vanderbilt University, Nashville, TN, 1984)
109. J.D. Schaffer, Multiple objective optimization with vector evaluated genetic algorithms, in *Proceedings of the 1st International Conference on Genetic Algorithms* (L. Erlbaum Associates Inc. Hillsdale, NJ, USA, 1985), pp. 93–100
110. C.M. Fonseca, P.J. Fleming, An overview of evolutionary algorithms in multiobjective optimization. J. Evol. Comput. **3**(1), 1–16 (1995)
111. M. Erickson, A. Mayer, J. Horn, Multi-objective optimal design of groundwater remediation systems: application of the niched pareto genetic algorithm (npga). Adv. Water Res. **25**(1), 51–65 (2002)
112. N. Srinivas, K. Deb, Multiobjective optimization using nondominated sorting in genetic algorithms. J. Evol. Comput. **2**(3), 221–248 (1994)
113. E. Zitzler, L. Thiele, Multiobjective evolutionary algorithms: a comparative case study and the strength Pareto approach. IEEE Trans. Evol. Comput. **3**(4), 257–271 (1999)
114. E. Zitzler, M. Laumanns, L. Thiele, *SPEA2: Improving the Strength Pareto Evolutionary Algorithm* (Technical Report, Swiss Federal Institute of Techonology, Zurich, Switzerland, 2001)
115. J.D. Knowles, D.W. Corne, Approximating the nondominated front using the Pareto archived evolution strategy. Evol. Comput. **8**(2), 149–172 (2000)
116. D.W. Corne, N.R. Jerram, J. Knowles, J. Oates, PESA-II: region-based selection in evolutionary multiobjective optimization, in *Proceedings of the Genetic and Evolutionary Computation Conference (GECCO-2001)* (Morgan Kaufmann, San Francisco, CA, 2001)
117. A.D. Brink, Minimum spatial entropy threshold selection. IEEE Proc. Vis. Image Sig. Process. **142**(3), 128–132 (1995)
118. O.J. Tobias, R. Seara, Image segmentation by histogram thresholding using fuzzy sets. IEEE Trans. Image Process. **11**(12), 1457–1465 (2002)
119. X. Li, Z. Zhao, H.D. Cheng, Fuzzy entropy threshold approach to breast cancer detection. Inf. Sci. Appl. **4**(1), 49–56 (1995)
120. A.Z. Arifin, A. Asano, Image segmentation by histogram thresholding using hierarchical cluster analysis. Pattern Recogn. Lett. **27**(13), 1515–1521 (2006)

121. N. Otsu, A threshold selection method from gray-level histograms. IEEE Trans. Syst. Man Cybern. **9**(1), 62–66 (1979)

122. S.H. Kwon, Threshold selection based on cluster analysis. Pattern Recogn. Lett. **25**(9), 1045–1050 (2004)

123. N. Vandenbroucke, L. Macaire, J.G. Postaire, Color image segmentation by pixel classification in an adapted hybrid color space: application to soccer image analysis. Comput. Vis. Image Underst. **90**(2), 190–216 (2003)

124. K.S. Tan, N.A.M. Isa, Color image segmentation using histogram thresholding-fuzzy C-means hybrid approach. Pattern Recogn. **44**(1), 1–15 (2011)

125. B. Sumengen, B.S. Manjunath, Multi-scale edge detection and image segmentation, in *Proceedings of European Signal Processing Conference (EUSIPCO)* (2005)

126. M. Tabb, N. Ahuja, Multiscale image segmentation by integrated edge and region detection. IEEE Trans. Image Process. **6**(5), 642–655 (1997)

127. J. Fan, D.K.Y. Yau, A.K. Elmagarmid, W.G. Aref, Automatic image segmentation by integrating color-edge extraction and seeded region growing. IEEE Trans. Image Process. **10**(10), 1454–1466 (2001)

128. S. De, S. Bhattacharyya, P. Dutta, Optimized multilevel image segmentation: a comparative study, in *Proceedings of IEEE National Conference on Computing and Communication Systems (CoCoSys-09)* (2009), pp. 200–205

129. S.A. Hijjatoleslami, J. Kittler, Region growing: a new approach. IEEE Trans. Image Process. **7**(7), 1079–1084 (1998)

130. J.C. Pichel, D.E. Singh, F.F. Rivera, Image segmentation based on merging of sub-optimal segmentations. Pattern Recogn. Lett. **27**(10), 1105–1116 (2006)

131. R. Adams, L. Bischof, Seeded region growing. IEEE Trans. Pattern Anal. Mach. Intell. **16**(6), 641–647 (1994)

132. F.Y. Shih, S. Cheng, Automatic seeded region growing for color image segmentation. Image Vis. Comput. **23**(10), 877–886 (2005)

133. O. Gmez, J.A. Gonzlez, E.F. Morales, Image segmentation using automatic seeded region growing and instance-based learning, in *Proceedings of CIARP'07* (2007), pp. 192–201

134. S. De, S. Bhattacharyya, P. Dutta, Multilevel image segmentation using variable threshold based OptiMUSIG activation function, in *Proceedings of National Conference on Nanotechnology and its Application in Quantum Computing (NAQC 2008)* (2008), pp. 44–51

135. J. Lie, M. Lysaker, X. Tai, A variant of the level set method and applications to image segmentation. Math. Comput. **75**, 1155–1174 (2006)

136. S. De, S. Bhattacharyya, P. Dutta, Multilevel image segmentation using OptiMUSIG activation function with fixed and variable thresholding: a comparative study, in *Applications of Soft Computing: From Theory to Praxis, Advances in Intelligent and Soft Computing*, eds. by J. Mehnen, M. Koppen, A. Saad, A. Tiwari (Berlin, Heidelberg) (Springer, 2009), pp. 53–62

137. G. Guo, S. Ma, Bayesian learning, global competition, and unsupervised image segmentation, in *Proceedings of Fourth International Conference on Signal Processing* (1998), pp. 986–989

138. J. Serra, *Image Analysis and Mathematical Morphology* (Academic Press, 1982)

139. J.M. Gauch, Image segmentation and analysis via multiscale gradient watershed hierarchies. IEEE Trans. Image Process. **8**(1), 69–79 (1999)

140. L.J. Belaid, W. Mourou, Image segmentation: a watershed transformation algorithm. Image Anal. Stereol. **28**(2), 93–102 (2009)

141. S. Mukhopadhyay, B. Chanda, Multiscale morphological segmentation of gray-scale images. IEEE Trans. Image Process. **12**(5), 533–549 (2003)

142. O. Lzoray, C. Charrier, Color image segmentation using morphological clustering and fusion with automatic scale selection. Pattern Recogn. Lett. **30**(4), 397–406 (2009)

143. J. Shi, J. Malik, Normalized cuts and image segmentation. IEEE Trans. Pattern Anal. Mach. Intell. **22**(8), 888–905 (2000)

144. X. Liu, D. Wang, Image and texture segmentation using spectral histograms. IEEE Trans. Image Process. **15**(10), 3066–3077 (2006)

145. W. Yang, L. Guo, T. Zhao, G. Xiao, Improving watersheds image segmentation method with graph theory, in *2nd IEEE Conference on Industrial Electronics and Applications* (2007), pp. 2550–2553

146. D. Comaniciu, P. Meer, Mean shift: a robust approach toward feature space analysis. IEEE Trans. Pattern Anal. Mach. Intell. **24**(5), 603–619 (2002)

147. Q. Luo, T.M. Khoshgoftaar, Unsupervised multiscale color image segmentation based on mdl principle. IEEE Trans. Image Process. **15**(9), 2755–2761 (2006)

148. T. Wenbing, J. Hai, Z. Yimin, Color image segmentation based on mean shift and normalized cuts. IEEE Trans. Syst. Man Cybern. Part B. Cybern. **37**(5), 1382–1389 (2007)

149. S. Wang, J.M. Siskind, Image segmentation with ratio cut. IEEE Trans. Pattern Anal. Mach. Intell. **25**(6), 675–690 (2003)

150. S. Makrogiannis, G. Economou, S. Fotopoulos, N.G. Bourbakis, Segmentation of color images using multiscale clustering and graph theoretic region synthesis. IEEE Trans. Syst. Man. Cybern. Part A **35**, 224–238 (2005)

151. P. Salembier, L. Garrido, Binary partition tree as an efficient representation for image processing, segmentation, and information retrieval. IEEE Trans. Image Process. **9**(4), 561–576 (2000)

152. J. Malik, S. Belongie, T. Leung, J. Shi, Contour and texture analysis for image segmentation. Int. J. Comput. Vis. **43**(1), 7–27 (2001)

153. J.T. Tou, R.C. Gonzalez, *Pattern Recognition* (U.K., Principles (Addison-Wesley, London, 1974)

154. C. Rosenberger, K. Chehdi, Unsupervised clustering method with optimal estimation of the number of clusters: application to image segmentation, in *15th International Conference on Pattern Recognition*, vol. 1, (Barcelona, Spain, 2000), pp. 656–659

155. T. Kanungo, D.M. Mount, N.S. Netanyahu, C.D. Piatko, R. Silverman, A.Y. Wu, An efficient k-means clustering algorithm: analysis and implementation. IEEE Trans. Pattern Anal. Mach. Intell. **24**(7), 881–892 (2002)

156. S.H. Choi, P. Rockett, The training of neural classifiers with condensed datasets. IEEE Trans. Syst. Man Cybern. Part B: Cybern. **32**(2), 202–206 (2002)

157. Y. Jiang, Z. Zhou, Som ensemble-based image segmentation. Neural Process. Lett. **20**(3), 171–178 (2004)

158. R. Freeman, H. Yin, N.M. Allinson, Self-organising maps for tree view based hierarchical document clustering. IEEE Int. Joint Conf. Neural Netw. (IJCNN'02) **2**, 1906–1911 (2002)

159. P.L. Chang, W.G. Teng, Exploiting the self-organizing map for medical image segmentation, in *Twentieth IEEE International Symposium on Computer-Based Medical Systems* (2007), pp. 281–288

160. W.E. Reddick et al., Automated segmentation and classification of multispectral magnetic resonance images of brain using artificial neural networks. IEEE Trans. Med. Imaging **16**(6), 911–918 (1997)

161. M.N. Nasrabadi, W. Li, Object recognition by a Hopfield neural network. IEEE Trans. Syst. Man Cybern. **21**(6), 1523–1535 (1991)

162. S. Munshi, S. Bhattacharyya, A.K. Datta, Photonic implementation of Hopfield neural network for associative pattern recognition. Proc. Int. Conf. Fiber Opt. Photon. Proc. SPIE **4417**, 558–562 (2001)

163. K.S. Cheng, J.S. Lin, C.W. Mao, The application of competitive Hopfield neural network to medical image segmentation. IEEE Trans. Med. Imaging **15**(4), 560–567 (1996)

164. B. Kosko, *Neural Networks and Fuzzy Systems: A Dynamical Systems Approach to Machine Intelligence* (Prentice-Hall, Englewood Cliffs, 1992)

165. A. Ghosh, S.K. Pal, Neural network, self-organization and object extraction. Pattern Recogn. Lett. **13**(5), 387–397 (1992)

166. K. Fukushima, Neocognitron: a self-organizing multilayer neural network model for a mechanism of pattern recognition unaffected by shift in position. Biol. Cybern. **36**, 193–202 (1980)

167. S. Amari, *Organization of Neural Networks: Structures and Models (Mathematical theory of self-organization in neural nets)* (Academic Press, New York, 1988)

168. S.H. Park, I.D. Yun, S.U. Lee, Color image segmentation based on 3-D clustering: morphological approach. Pattern Recogn. **31**(8), 1061–1076 (1998)
169. T.D. Pham, Image segmentation using probabilistic fuzzy C-means clustering. Proc. IEEE Int. Conf. Image Process. **1**, 722–725 (2001)
170. D. Chi, Self-Organizing Map-Based Color Image Segmentation with k-Means Clustering and Saliency Map. ISRN Signal Processing Article ID 393891 (2011). doi:10.5402/2011/393891
171. W. Lee, W. Kim, Self-organization neural network for multiple texture image segmentation. Proc. IEEE Tencon **1**, 730–733 (1999)
172. S. De, S. Bhattacharyya, P. Dutta, Efficient grey-level image segmentation using an optimised MUSIG (OptiMUSIG) activation function. Int. J. Parallel Emerg. Distrib. Syst. **26**(1), 1–39 (2010)
173. A. Ghosh, A. Sen, *Soft Computing Approach to Pattern Recognition and Image Processing* (World Scientific, 2002)
174. S. Bhattacharyya, U. Maulik, P. Dutta, Multilevel image segmentation with adaptive image context based thresholding. Appl. Soft Comput. **11**(1), 946–962 (2011)
175. V.V. Vinod, S. Chaudhury, J. Mukherjee, S. Ghosh, A connectionist approach for clustering with applications in image analysis. IEEE Trans. Syst. Man Cybern. **24**(3), 365–384 (1994)
176. M.N. Ahmed, A.A. Farag, Two-stage neural network for volume segmentation of medical images. Pattern Recogn. Lett. **18**, 1143–1151 (1997)
177. S. De, S. Bhattacharyya, P. Dutta, OptiMUSIG : an optimized gray level image segmentor, in *Proceedings of 16th International Conference on Advanced Computing and Communications (ADCOM 2008)* (2008), pp. 78–87
178. J. Zhang, Q. Liu, Z. Chen, A medical image segmentation method based on SOM and wavelet transforms. J. Commun. Comput. **2**(5), 46–50 (2005)
179. J. Jiang, P. Trundle, J. Ren, Medical image analysis with artificial neural networks. Comput. Med. Imaging Graph. **34**(8), 617–631 (2010)
180. N. Torbati, A. Ayatollahi, A. Kermani, An efficient neural network based method for medical image segmentation. Comput. Biol. Med. **44**, 76–87 (2014)
181. J. Alirezaie, M.E. Jernigan, C. Nahmias, Automatic segmentation of cerebral MR images using artificial neural networks. IEEE Trans. Nucl. Sci. **45**(4), 2174–2182 (1998)
182. M. Sammouda, R. Sammouda, N. Niki, M. Benaichouche, Tissue color images segmentation using artificial neural networks. Proceedings of IEEE International Symposium on Biomedical Imaging: Nano to Macro **1**, 145–148 (2004)
183. T. Uchiyama, M.A. Arbib, Color image segmentation using competitive learning. IEEE Trans. Pattern Anal. Mach. Intell. **16**(12), 1197–1206 (1994)
184. J.H. Wang, J.D. Rau, W.J. Liu, Two-stage clustering via neural networks. IEEE Trans. Neural Netw. **14**(3), 606–615 (2003)
185. N.C. Yeo, K.H. Lee, Y.V. Venkatesh, S.H. Ong, Colour image segmentation using the self-organizing map and adaptive resonance theory. Image Vis. Comput. **23**(12), 1060–1079 (2005)
186. Y. Jiang, K.J. Chen, Z.H. Zhou, Som-based image segmentation, in *Proceedings of 9th Conference on Rough Sets, Fuzzy Sets, Data Mining and Granular Computing* (2003), pp. 640–643
187. J. Vesanto, E. Alhoniemi, Clustering of the self-organizing map. IEEE Trans. Pattern Anal. Mach. Intell. **11**(3), 586–600 (2000)
188. S.H. Ong, N.C. Yeo, K.H. Lee, Y.V. Venkatesh, D.M. Cao, Segmentation of color images using a two-stage self-organizing network. Image Vis. Comput. **20**(4), 279–289 (2002)
189. G. Dong, M. Xie, Color clustering and learning for image segmentation based on neural networks. IEEE Trans. Neural Netw. **16**(4), 925–936 (2005)
190. A.R.F. Arajo, D.C. Costa, Local adaptive receptive field self-organizing map for image color segmentation. Image Vis. Comput. **27**(9), 1229–1239 (2009)
191. S.B. Park, J.W. Lee, S.K. Kim, Content-based image classification using a neural network. Pattern Recogn. Lett. **25**(3), 287–300 (2004)
192. T. Qian, M. Li, Multispectral mr images segmentation using som network, in *Proceedings of Fourth International Conference on Computer and Information Technology (CIT'04)* (2004), pp. 155–158

193. S. De, S. Bhattacharyya, S. Chakraborty, Color image segmentation using parallel OptiMUSIG activation function. Appl. Soft Comput. **12**(10), 3228–3236 (2012)
194. Y. Sirisathitkul, S. Auwatanamongkol, B. Uyyanonvara, Color image quantization using distances between adjacent colors along the color axis with highest color variance. Pattern Recogn. Lett. **25**(3), 1025–1043 (2004)
195. S. Bhattacharyya, P. Dutta, U. Maulik, P.K. Nandi, Multilevel activation functions for true color image segmentation using a self supervised parallel self organizing neural network (PSONN) architecture: A comparative study. Int. J. Comput. Sci. **2**(1), 09–21 (2007)
196. S. Bhattacharyya, K. Dasgupta, Color object extraction from a noisy background using parallel multilayer self-organizing neural networks, in *Proceedings of CSI-YITPA(E) 2003* (2003), pp. 32–36
197. S. Bhattacharyya, A brief survey of color image preprocessing and segmentation techniques. Pattern Recongn. Res. **6**(1), 120–129 (2011)
198. H.D. Cheng, X.H. Jiang, Y. Sun, J. Wang, Color image segmentation: advances and prospects. Pattern Recogn. **34**(12), 2259–2281 (2001)
199. J.C. Bezdek, *Pattern Recognition with Fuzzy Objective Function Algorithms* (Plenum Press, New York, 1981)
200. M.N. Ahmed, S.M. Yamany, N. Mohamed, A.A. Farag, T. Moriarty, A modified fuzzy c-means algorithm for bias field estimation and segmentation of MRI data. IEEE Trans. Med. Imaging **21**(3), 193–199 (2002)
201. J.C. Noordam, W.H.A.M. van den Broek, L.M.C. Buydens, Geometrically guided fuzzy C-means clustering for multivariate image segmentation. Proc. Int. Conf. Pattern Recogn. **1**, 32–36 (2000)
202. Y.A. Tolias, S.M. Panas, Image segmentation by a fuzzy clustering algorithm using adaptive spatially constrained membership functions. IEEE Trans. Syst. Man Cybern Part A: Syst. Humans **28**(3), 359–369 (1998)
203. A.W.C. Liew, H. Yan, N.-F. Law, Image segmentation based on adaptive cluster prototype estimation. IEEE Trans. Fuzzy Syst. **13**(4), 444–453 (2005)
204. S. Chen, D. Zhang, Robust image segmentation using fcm with spatial constraints based on new kernel-induced distance measure. IEEE Trans. Syst. Man. Cybern. Part B: Cybern. **34**(4), 1907–1916 (2004)
205. M.R. Rezaee, P.M.J. Zwet, B.P.F. Lelieveldt, R.J. Geest, J.H.C. Reiber, A multiresolution image segmentation technique based on pyramidal segmentation and fuzzy clustering. IEEE Trans. Image Process. **9**(7), 1238–1248 (2000)
206. A.S. Pednekar, I.A. Kakadiaris, Image segmentation based on fuzzy connectedness using dynamic weights. IEEE Trans. Image Process. **15**(6), 1555–1562 (2006)
207. E.E. Kerre, M. Nachtegael (eds.), Fuzzy techniques in image processing, in *Studies in Fuzziness and Soft Computing* vol. 52 (2000), pp. 337–369
208. S. Makrogiannis, I. Vanhamel, H. Sahli, S. Fotopoulos, Scale space segmentation of color images using watersheds and fuzzy region merging. Proc. Int. Conf. Image Process. **1**, 734–737 (2001)
209. S. Makrogiannis, G. Economou, S. Fotopoulos, Region oriented compression of color images using fuzzy inference and fast merging. Pattern Recogn. **35**(9), 1807–1820 (2002)
210. S.B. Chaabane, M. Sayadi, F. Fnaiech, E. Brassart, Colour image segmentation using homogeneity method and data fusion techniques. EURASIP J. Adv. Sig. Process. Article ID 367297, 11 pages (2010). doi:10.1155/2010/367297
211. Y. Yang, C. Zheng, P. Lin, Image thresholding based on spatially weighted fuzzy C-means clustering," in *Proceedings of the 4th International Conference on Computer and Information Technology (CIT'04)* (2004), pp. 184–189
212. P.A. Estevez, R.J. Flores, C.A. Perez, Color image segmentation using fuzzy min-max neural networks. Proc. IEEE Int. Joint Conf. Neural Netw. **5**, 3052–3057 (2005)
213. C.C. Kang, W.J. Wang, Fuzzy based seeded region growing for image segmentation, in *2009 Annual Meeting of the North American Fuzzy Information Processing Society* (2009), pp. 1–5

214. J. Yang, S. Hao, P. Chung, Color image segmentation using fuzzy C-means and eigenspace projections. Sig. Process. **82**(3), 461–472 (2002)
215. A.S. Frisch, Unsupervised construction of fuzzy measures through self-organizing feature maps and its application in color image segmentation. Int. J. Approximate Reasoning **41**(1), 23–42 (2006)
216. H.H. Muhammed, Unsupervised fuzzy clustering and image segmentation using weighted neural networks, in *Proceedings of the 12th International Conference on Image Analysis and Processing (ICIAP'03)* (2003), pp. 308–313
217. D.E. Goldberg, *Genetic Algorithm in Search Optimization and Machine Learning* (Addison-Wesley, New York, 1989)
218. M. Mitchell, *An Introduction to Genetic Algorithms* (MIT Press, Cambridge, MA, 1996)
219. B. Bhanu, S. Lee, J. Ming, Adaptive image segmentation using a genetic algorithm. IEEE Trans. Syst. Man Cybern. **25**(12), 1543–1567 (1995)
220. J.T. Alander, *Indexed Bibliography of Genetic Algorithms in Optics and Image Processing, Technical Report 94-1-OPTICS* (University of Vaasa, Vaasa, Finland, Department of Information Technology and Production Economics, 2000)
221. S. Bandyopadhyay, C.A. Murthy, S.K. Pal, Pattern classification with genetic algorithms. Pattern Recogn. Lett. **16**(5), 801–808 (1995)
222. C.A. Murthy, N. Chowdhury, In search of optimal clusters using genetic algorithm. Pattern Recogn. Lett. **17**(8), 825–832 (1996)
223. M. Yoshimura, S. Oe, Evolutionary segmentation of texture using genetic algorithms towards automatic decision of optimum number of segmentation areas. Pattern Recogn. **32**(12), 2041–2054 (1999)
224. C.-T. Li, R. Chiao, Multiresolution genetic clustering algorithm for texture segmentation. Image Vis. Comput. **21**(11), 955–966 (2003)
225. C.C. Lai, C.Y. Chang, A hierarchical genetic algorithm based approach for image segmentation. Proc. IEEE Int. Conf. Netw. Sens. Control **2**, 1284–1288 (2004)
226. X. Jin, C.H. Davis, A genetic image segmentation algorithm with a fuzzy-based evaluation function. Proc. IEEE Int. Conf. Fuzzy Syst. **2**, 938–943 (2003)
227. S.K. Pal, S. Bandyopadhyay, C.A. Murthy, Genetic algorithms for generation of class boundaries. IEEE Trans. Syst. Man Cyber. Part B: Cybern. **28**(8), 816–827 (1998)
228. W.B. Tao, J.W. Tian, J. Liu, Image segmentation by three-level thresholding based on maximum fuzzy entropy and genetic algorithm. Pattern Recogn. Lett. **24**(16), 3069–3078 (2003)
229. H.D. Cheng, Y.H. Chen, X.H. Jiang, Thresholding using two-dimensional histogram and fuzzy entropy principle. IEEE Trans. Image Process. **9**(4), 732–735 (2000)
230. M.S. Zhao, A.M.N. Fu, H. Yan, A technique of three level thresholding based on probability partition and fuzzy 3-partition. IEEE Trans. Fuzzy Syst. **9**(3), 469–479 (2001)
231. L. Ballerini, L. Bocchi, C. Johansson, Image segmentation by a genetic fuzzy c-means algorithm using color and spatial information. Appl. Evol. Comput. **3005**, 260–269 (2004)
232. M. Haseyama, M. Kumagai, and H. Kitajima, A genetic algorithm based image segmentation for image analysis, in *Proceedings of IEEE International Conference on Acoustics, Speech, and Signal Processing* (1999), pp. 3445–3448
233. F. Yang, T. Jiang, Pixon-based image segmentation with Markov random fields. IEEE Trans. Image Process. **12**(12), 1552–1559 (2003)
234. R.Q. Feitosa, G.A. O.P. Costa, T.B. Cazes, A genetic approach for the automatic adaptation of segmentation parameters, in *Proceedings of OBIA06* (2006)
235. H. Zhu, O. Basir, Image segmentation with GA optimized fuzzy reasoning. IEEE Int. Conf. Fuzzy Syst. **2**, 990–995 (2003)
236. M. Yu, N. Eua-anant, A. Saudagar, L. Udpa, Genetic algorithm approach to image segmentation using morphological operations. Proc. Int. Conf. Image Process. **3**, 775–779 (1998)
237. C. Sun, H. Chao, Y. Sun, Genetic-based clustering neural networks and applications. IEEE Int. Conf. Intell. Process. Syst. **1**, 439–443 (1997)
238. K. Hammouche, M. Diaf, P. Siarry, A multilevel automatic thresholding method based on a genetic algorithm for a fast image segmentation. Comput. Vis. Image Underst. **109**(2), 163–175 (2008)

239. U. Maulik, Medical image segmentation using genetic algorithms. IEEE Trans. Inf. Technol. Biomed. **13**(2), 166–173 (2009)
240. F. Zhao, X. Xie, An overview of interactive medical image segmentation. Br. Mach. Vis. Assoc. Soc. Pattern Recogn. **2013**(7), 1–22 (2013)
241. C.C. Lai, C.Y. Chang, A hierarchical evolutionary algorithm for automatic medical image segmentation. Expert Syst. Appl. **36**(1), 248–259 (2007)
242. S. De, S. Bhattacharyya, S. Chakraborty, Application of pixel intensity based medical image segmentation using NSGA II based OptiMUSIG activation function, in *2014 International Conference on Computational Intelligence and Communication Networks (CICN2014)* (2014), pp. 262–267
243. K.K. Pavan, V.S. Srinivas, A. SriKrishna, B.E. Reddy, Automatic tissue segmentation in medical images using differential evolution. J. Appl. Sci. **12**(6), 587–592 (2012)
244. P. Ghosh, M. Mitchell, Segmentation of medical images using a genetic algorithm, in *Proceedings of the 8th Annual Conference on Genetic and Evolutionary Computation* (2006), pp. 1171–1178
245. E.A. Zanaty, A.S. Ghiduk, A novel approach based on genetic algorithms and region growing for magnetic resonance image (MRI) segmentation. Comput. Sci. Inf. Syst. **10**(3), 1319–1342 (2013)
246. N. Benamrane, A. Fekir, Medical images segmentation by neuro-genetic approach, in *Proceedings of the Ninth International Conference on Information Visualisation* (2005), pp. 981–986
247. M.E. Farmer, D. Shugars, Application of genetic algorithms for wrapper-based image segmentation and classification, *IEEE Congress on Evolutionary Computation* (2006), pp. 1300–1307
248. P. Zingaretti, G. Tascini, L. Regini, Optimising the colour image segmentation, in *Proceedings of VIII Convegno dell Associazione Italiana per lIntelligenza Artificiale* (2002)
249. M. Gong, Y.H. Yang, Genetic-based multiresolution color image segmentation. Proc. Vis. Interface **2001**, 141–148 (2001)
250. G. Pignalberi, R. Cu-cchiara, L. inque, S. Levialdi, Tuning range image segmentation by genetic algorithm. EURASIP J. Appl. Sig. Process. **8**, 780–790 (2003)
251. J.J. Hopfield, D.W. Tank, Neural computation of decisions in optimization of problems. Biol. Cybern. **52**, 141–152 (1985)
252. S.C. Ngan, X. Hu, Analysis of functional magnetic resonance imaging data using self-organizing mapping with spatial connectivity. Magn. Reson. Med. **41**(5), 939–946 (1999)
253. http://www.imaios.com/en/e-Anatomy/Head-and-Neck/Brain-MRI-in-axial-slices
254. http://www.imaios.com/en/e-Anatomy/Head-and-Neck/Brain-MRI-3D
255. J. Liu, Y. Yang, Multiresolution color image segmentation. IEEE Trans. Pattern Anal. Mach. Intell. **16**, 689–700 (1994)
256. M. Borsotti, P. Campadelli, R. Schettini, Quantitative evaluation of color image segmentation results. Pattern Recogn. Lett. **19**, 741–747 (1998)
257. Y. Zhang, A survey on evaluation methods for image segmentation. Pattern Recogn. **29**(8), 1335–1346 (1996)
258. S. De, S. Bhattacharyya, S. Chakraborty, Efficient color image segmentation by a parallel optimized (ParaOptiMUSIG) activation function, in *Global Trends in Intelligent Computing Research and Development*, ed. by B.K. Tripathy, D.P. Acharjya (IGI Global, Hershey, PA, USA, 2013), pp. 19–50
259. L. Lucchese, S. Mitra, Color image segmentation: a state-of-the-art survey, in *Proceedings of The Indian National Science Academy (INSA-A)*, vol. 67, (New Delhi, India), Image Processing, Vision, and Pattern Recognition (2001), pp. 207–221
260. Y. Sun, G. He, Segmentation of high-resolution remote sensing image based on marker-based watershed algorithm. Proceedings of Fifth International Conference on Fuzzy Systems and Knowledge Discovery **4**, 271–276 (2008)
261. N. Ikonomakis, K.N. Plataniotis, A.N. Venetsanopoulos, Color image segmentation for multimedia applications. J. Intell. Robot. Syst. **28**(1–2), 5–20 (2000)

262. E. Lpez-Ornelas, High resolution images: segmenting, extracting information and GIS integration. World Acad. Sci. Eng. Technol. **3**(6), 150–155 (2009)
263. Y. Rui, T.S. Huang, Image retrieval: current techniques, promising directions and open issues. J. Vis. Commun. Image Represent. **10**(1), 39–62 (1999)
264. H.C. Chen, W.J. Chien, S.J. Wang, Contrast-based color image segmentation. IEEE Sig. Process. Lett. **11**(7), 641–644 (2004)
265. D.W. Jacobs, D. Weinshall, Y. Gdalyahu, Classification with nonmetric distances: Image retrieval and class representation. IEEE Trans. Pattern Anal. Mach. Intell. **22**(6), 583–600 (2000)
266. D. Androutsos, K.N. Plataniotis, A.N. Venetsanopoulos, Distance measures for color image retrieval. Proc. IEEE Conf. Image Process. **2**, 770–774 (1998)
267. Y. Deng, B.S. Manjunath, H. Shin, Color image segmentation, in *Proceedings of IEEE Conference in Computer Vision Pattern Recognition* (1999), pp. 1021–1025
268. M.A. Ruzon, C. Tomasi, Color edge detection with the compass operator. Proc. IEEE Conf. Comput. Vis. Pattern Recogn. **2**, 160–166 (1999)
269. Y. Rubner, C. Tomasi, L.J. Guibas, A metric for distributions with applications to image databases, in *Proceedings of International Conference on Computer Vision* (1998), pp. 59–66
270. B.A. Maxwell, S.J. Brubaker, Texture edge detection using the compass operator, in *Proceedings of the British Machine Conference*, ed. by R. Harvey, A. Bangham (2003), pp. 56.1–56.10
271. W. Cai, J. Wu, A.C.S. Chung, Shape-based image segmentation using normalized cuts, in *IEEE International Conference on Image Processing* (2006), pp. 1101–1104
272. S. Wang, T. Kubota, T. Richardson, Shape correspondence through landmark sliding, in *Proceedings of the 2004 IEEE Computer Society Conference on Computer Vision and Pattern Recognition (CVPR 2004)*, vol. 1 (2004), pp. 143–150
273. S. De, S. Bhattacharyya, S. Chakraborty, True color image segmentation by an optimized multilevel activation function, in *Proceedings of 2010 IEEE International Conference on Computational Intelligence and Computing Research (2010 IEEE ICCIC)* (2010), pp. 545–548
274. M. Ceccarelli, A. Petrosino, Multi-feature adaptive segmentation for SAR image. J. Neurocomput. **14**, 345–363 (1997)
275. M.R. Azimi-Sadjadi, S. Ghaloum, R. Zoughi, Terrain classification in SAR images using principal components analysis and neural networks. IEEE Int. Joint Conf. Neural Netw. **3**, 2390–2395 (1991)
276. C.A.C. Coello, A.D. Christiansen, An approach to multiobjective optimization using genetic algorithms, in *Intelligent Engineering Systems Through Artificial Neural Networks*, ed. by C.H. Dagli, M. Akay, C.L. P.C. B.R. Fernndez, J. Ghosh, vol. 5 (ASME Press, St. Louis Missouri, USA, 1995), pp. 411–416. Fuzzy Logic and Evolutionary Programming
277. D.A.V. Veldhuizen, G.B. Lamont, Multiobjective evolutionary algorithms: analyzing the state-of-the-art. Evol. Comput. **8**(2), 125–147 (2000)
278. B. Chin-Wei, M. Rajeswari, Multiobjective optimization approaches in image segmentation-the directions and challenges. Int. J. Adv. Soft Comput. Appl. **2**(1), 40–64 (2010)
279. S. Shirakawa, T. Nagao, Evolutionary image segmentation based on multiobjective clustering, *Congress on Evolutionary Computation (CEC '09)* (2009), pp. 2466–2473
280. A. Nakib, H. Oulhadj, P. Siarry, Image thresholding based on Pareto multiobjective optimization. Eng. Appl. Artif. Intell. **23**(3), 313–320 (2010)
281. D.L. Davies, D.W. Bouldin, A cluster separation measure. IEEE Trans. Pattern Recogn. Mach. Intell. **1**(2), 224–227 (1979)
282. M. Halkidi, M. Vazirgiannis, Clustering validity assessment using multi representatives, in *Proceedings of the Hellenic Conference on Artificial Intelligence (SETN '02)* (Thessaloniki, Greece, 2002), pp. 32–36
283. M. Sezgin, O.K. Ersoy, B. Yazgan, Segmentation of remote sensing images using multistage unsupervised learning. Appl. Dig. Image Process. XXVII. Proc. SPIE **616–623**, 32–36 (2004)
284. A.R.S. Marcal, J.S. Borges, Estimation of the "natural" number of classes of a multispectral image, in *IEEE International Geoscience and Remote Sensing Symposium (IGARSS '05)*, vol. 6 (2005), pp. 3788–3791

285. N.R. Pal, J. Biswas, Cluster validation using graph theoretic concepts. Pattern Recogn. **30**(6), 847–857 (1997)
286. H. Zhang, J. Fritts, S. Goldman, An entropy-based objective evaluation method for image segmentation, in *Proceedings of SPIE Storage and Retrieval Methods and Applications for Multimedia* (2004), pp. 38–49
287. S. De, S. Bhattacharyya, S. Chakraborty, Multilevel and color image segmentation by NSGA II based OptiMUSIG activation function, in *Handbook of Research on Advanced Hybrid Intelligent Techniques and Applications*, ed. by S. Bhattacharyya, P. Banerjee, D. Majumdar, P. Dutta (IGI Global, Hershey, PA, USA, 2015), pp. 321–348
288. U. Maulik, S. Bandyopadhyay, A. Mukhopadhyay, *Multiobjective Genetic Algorithm-Based Fuzzy Clustering* (Springer, Berlin, 2011)
289. S. Bandyopadhyay, S. Saha, *Some single and multiobjective optimization techniques, in Unsupervised Classification* (Springer, Berlin, 2013), pp. 17–58
290. Z. Huang, Y. Fang, Z. Zhen, C. Zhang, Multi-objective fuzzy clustering method for image segmentation based on variable-length intelligent optimization algorithm, in Advances in Computation and Intelligence, vol. 6382 of *Lecture Notes in Computer Science* (Springer, Berlin, 2010), pp. 329–337
291. A. Nakib, H. Oulhadj, P. Siarry, Image histogram thresholding based on multiobjective optimization. Sig. Process. **87**, 2516–2534 (2007)
292. S. Bandyopadhyay, A.K. Srivastava, S.K. Pal (eds.) *Multi-objective Variable String Genetic Classifier: Application of Remote Sensing Image* (World Scientific, 2002)
293. S. De, S. Bhattacharyya, S. Chakraborty, Color image segmentation by NSGA-II based ParaOptiMUSIG activation function, in *2013 International Conference on Machine Intelligence Research and Advancement (ICMIRA-2013)*, December 2013, pp. 105–109
294. F. Hoppner, F. Klawonn, R. Kruse, T. Runkler, *Fuzzy cluster analysis-methods for classification, Data Analysis and Image Recognition* (Plenum Press, New York, 1999)
295. A.K. Jain, R. Duin, J. Mao, Statistical pattern recognition: a review. IEEE Trans. Pattern Anal. Mach. Intell. **22**(1), 4–37 (2000)
296. G. Hamerly, C. Elkan, Learning the k in k-means, in *Seventh Annual Conference on Neural Information Processing Systems (NIPS)* (2003), pp. 281–288
297. J. Hartigan, M. Wang, A K-means clustering algorithm. Appl. Stat. **28**, 100–108 (1979)
298. D. Pelleg, A. Moore, X-means: extending k-means with efficient estimation of the number of clusters, in *Proceedings of 17th International Conference Machine Learning*, (Stanford, CA) (2000), pp. 727–734
299. Z. Huang, Extensions to the k-means algorithm for clustering large data sets with categorical values. Data Min. Knowl. discovery **2**(3), 283–304 (1998)
300. Y.M. Cheung, k-means: a new generalized k-means clustering algorithm. Pattern Recogn. Lett. **24**(15), 2883–2893 (2003)
301. Y. Yang, C. Zheng, P. Lin, Fuzzy c-means clustering algorithm with a novel penalty term for image segmentation. Opto-Electron. Rev. **13**(4), 309–315 (2005)
302. A. Lorette, X. Descombes, J. Zerubia, Fully unsupervised fuzzy clustering with entropy criterion. Int. Conf. Pattern Recogn. **3**, 3998–4001 (2000)
303. G.J. McLachlan, T. Krishnan, *The EM Algorithm and Extensions* (Wiley, 1997)
304. R.H. Sheikh, M.M. Raghuwanshi, A.N. Jaiswal, Genetic algorithm based clustering: a survey, in *First International Conference on Emerging Trends in Enggineering and Technology* (2008), pp. 314–319
305. V.V. Raghavan, K. Birchand, A clustering strategy based on a formalism of the reproductive process in a natural system, in *Proceedings of the Second International Conference on Information Storage and Retrieval* (1979), pp. 10–22
306. R. Srikanthan, R. George, N. Warsi, D. Prabhu, F.E. Petri, B.P. Buckles, A variable-length genetic algorithm for clustering and classification. Pattern Recogn. Lett. **16**, 789–800 (1995)
307. Y.C. Chiou, L.W. Lan, Theory and methodology genetic clustering algorithms. Eur. J. Oper. Res. **135**(2), 413–427 (2001)

308. V. Katari, S.C. Satapathy, J.V.R. Murthy, P.V.G.D.P. Reddy, Hybridized improved genetic algorithm with variable length chromosome for image clustering. Int. J. Comput. Sci. Netw. Secur. **7**, 121–131 (2007)

309. I. Sarafis, A.M.S. Zalzala, and P. Trinder, A genetic rule-based data clustering toolkit, in *Congress on Evolutionary Computation (CEC)* (2002), pp. 1238–1243

310. M.H.Y. Batistakis, M. Vazirgiannis, On clustering validation techniques. J. Intell. Inf. Syst. **17**, 107–145 (2001)

311. R.M. Othman, S. Deris, R.M. Illias, Z. Zakaria, S.M. Mohamad, Automatic clustering of gene ontology by genetic algorithm. Int. J. Inf. Technol. **3**(1), 37–46 (2005)

312. C. Y. Lee, E.K. Antonsson, Self-adapting vertices for mask-layout synthesis, in *Technical Proceedings of the 2000 International Conference on Modeling and Simulation of Microsystems* (2000), pp. 83–86

313. S. Bandyopadhyay, U. Maulik, Genetic clustering for automatic evolution of clusters and application to image classification. IEEE Trans. Syst. Man Cybern. Part B Cybern. **35**(6), 1197–1208 (2002)

314. J.C. Dunn, Well separated clusters and optimal fuzzy partitions. J. Cybern. **4**(1), 95–104 (1974)

315. J.C. Bezdek, Numerical taxonomy with fuzzy sets. J. Math. Biol. **1**(1), 57–71 (1974)

316. X.L. Xie, G. Beni, A validity measure for fuzzy clustering. IEEE Trans. Pattern Anal. Mach. Intell. **13**(8), 841–847 (1991)

317. M.K. Pakhira, S. Bandyopadhyay, U. Maulik, Validity index for crisp and fuzzy clusters. Pattern Recogn. Lett. **37**(3), 487–501 (2004)

318. C.H. Chou, M.C. Su, E. Lai, A new cluster validity measure and its application to image compression. Pattern Anal. Appl. **7**, 205–220 (2004)

319. S. De, S. Bhattacharyya, S. Chakraborty, Automatic data clustering by genetic algorithm validated by fuzzy intercluster hostility index, in *2014 Fourth International Conference of Emerging Applications of Information Technology (EAIT 2014)* (2014), pp. 58–63

320. S. De, S. Bhattacharyya, S. Chakraborty, A genetic algorithm based automatic image clustering technique validated by fuzzy intercluster hostility index, in *Proceedings of International Conference on Communication, Computers and Devices (ICCCD 2010)* (2010)

321. S. De, S. Bhattacharyya, P. Dutta, Automatic clustering by DE algorithm validated by fuzzy intercluster hostility index, in *Proceedings of International Conference on Emerging Trends in Computer Science, Communication and Information Technology (CSCIT2010)* (2010)

322. S. De, S. Bhattacharyya, P. Dutta, A differential evolution algorithm based automatic determination of optimal number of clusters validated by fuzzy intercluster hostility index, in *Proceedings of First IEEE International Conference on Advanced Computing (ICAC09)*(2009), pp. 105–111

323. S. De, S. Bhattacharyya, P. Dutta, Automatic image clustering by genetic algorithm validated by fuzzy intercluster hostility index and classical DE: a comparative study, in *Proceedings of National Conference on Ubiquitous Computing (NCUBQT 2009)* (2009), pp. 12–17

324. R.M. Haralick, L.G. Shapiro, Image segmentation techniques. Comput. Vis. Graph. Image Process. **29**(1), 100–132 (1985)

325. R. Rezaee, B.P.F. Lelieveldt, J.H.C. Reiber, A new cluster validity index for the fuzzy c-mean. Pattern Recogn. Lett. **19**(3–4), 237–246 (1998)

326. M. Halkidi, M. Vazirgiannis, Clustering validity assessment: finding the optimal partitioning of a data set, in *Proceedings of IEEE International Conference on Data Mining* (2001), pp. 187–194

327. S. Theodoridis, K. Koutroubas, *Pattern Recognition* (Academic Press, 1999)

328. C. Blake, E. Keough, C.J. Merz (1998). http://www.ics.uci.edu/~mlearn/MLrepository.html

Index

A
Activation function, 9
Activation state vector, 8
Ant Colony Optimisation, 4
Artificial intelligence, 1, 2, 19
Artificial Neural Network, 3, 8–10, 35
Automatic clustering differential evolution (ACDE), 40, 195, 196, 200, 201, 203, 205, 213, 214, 217

B
Backpropagation learning, 10
Boltzmann selection, 16

C
Chromosomes, 14, 16
Cluster validity indices, 198
Correlation coefficient, 51, 53
Crossover, 4, 14–19, 27, 38, 40, 51, 54, 98, 129, 142, 156
Crossover point, 16
Crossover probability, 16, 17, 98, 142, 156, 176
Crossover rate, 16
CS measure, 196, 198–200

D
DB index, 196, 198–200
Defuzzification interface, 12
Denormalisation, 13
Differential Evolution (DE), 18, 194, 195, 200, 201

E
Empirical goodness measures, 52

F
Feedback neural network, 10
Feedforward neural network, 10
Fitness function, 14
Fuzzification, 12
Fuzzy c-means (FCM), 37, 39, 43, 44, 194
Fuzzy inference engine, 12
Fuzzy Intercluster Hostility (FIH) index, 196, 197, 201–203, 205, 209, 213, 214, 217
Fuzzy logic, 2, 10–12, 33, 37
Fuzzy rule base/knowledge base, 12
Fuzzy rules, 12
Fuzzy set, 3, 10–13, 37
Fuzzy set theory, 12
Fuzzy system, 12

G
Gene pools, 14
Genetic Algorithm (GA), 4, 14–18, 25, 27, 28, 33, 38–40, 42–44, 53, 57, 85, 91, 92, 94, 101, 120, 194, 201
Genetic operators, 16

Dunn index, 196, 198

© Springer International Publishing AG 2016
S. De et al., *Hybrid Soft Computing for Multilevel Image and Data Segmentation*, Computational Intelligence Methods and Applications, DOI 10.1007/978-3-319-47524-0

Printed in the United States
By Bookmasters